T0137454

The Perception of Quality

George N. Kenyon · Kabir C. Sen

The Perception of Quality

Mapping Product and Service Quality to Consumer Perceptions

 Springer

George N. Kenyon
Kabir C. Sen
Lamar University
Beaumont, TX
USA

ISBN 978-1-4471-7040-2   ISBN 978-1-4471-6627-6 (eBook)
DOI 10.1007/978-1-4471-6627-6

Springer London Heidelberg New York Dordrecht

Printed on acid-free paper

Springer is part of Springer Science+Business Media (www.springer.com)

Contents

Chapter 1
What is Quality?

> *Quality is how we describe the value we perceive in the innate characteristics of a product or the attributes of a service.*
>
> *Quality Management is how we apply the theories, principles, and practices associated with the defining, creating, and delivering of products and services that generate value for our customers and society.*

The purpose of any business is to generate money. The reason for this objective is that without sufficient financial resources, the business cannot grow or maintain its assets, or develop new products or services, or hire and pay employees, or any other activity that society regards as necessary or worthwhile. But how does a business consistently generate money? The answer is simple: through the creation of value for customers, employees, investors, and society. Unfortunately, the consistent creation of value is not simple. Due to the interests of these four stakeholder groups being inextricably linked, sustained value can only occur when value is created for all four groups simultaneously.

So what is meant by value? For the customer, it involves the firm developing products and services that they consistently find useful or beneficial. For the employee, value is generated through being treated respectfully, being involved in decision making, engaging in meaningful work, and continued training and other developmental activities. For the investor, value is realized when the company delivers consistent high returns on their investments. For society, value is created when there is an increase in economic activity that is not accompanied by harmful environmental and/or social actions.

To consistently create value for customers, employees, investors, and society, companies must embrace a "total quality" philosophy and approach change from a holistic perspective. To be a world-class value creator, they will need to take quality management beyond the traditional understanding. Not only do they have to produce high-quality products and services in the traditional sense, but they must

© Springer-Verlag London 2015
G.N. Kenyon and K.C. Sen, *The Perception of Quality*,
DOI 10.1007/978-1-4471-6627-6_1

also produce high quality in a metaphysical sense. In other words, a commitment to quality should be inherently manifest in total creative process, from the design to the final delivery to the customer.

The word "quality" has its roots in the English language with Shakespeare using it in the phrase "the quality of mercy is not strained" in his play "The Merchant of Venice" written between 1596 and 1598. The dictionary defines quality by attributing two meanings to it. The first definition refers to the essential characteristics of a product; the second connotes a degree of excellence which a product has. For example, colloquial English often contrasts high-quality products with cheap, inferior imitations. However, "quality" is not synonymous with "expensive." A more realistic view of quality is a measure of a product's ability to live up to the average consumer's expectations about it. This approach to a definition of "quality" is related to the **concept of value** [1].

In order to create value, there must be a balance between the benefits provided by the product or service and the costs associated with consuming the service or using the product over its lifetime. It should be noted that all costs incurred by the customer might not be monetary. For example, given a choice between two computers from rival competitors, a consumer might find a negligible difference between their features and price, but could pick one that has a significantly earlier delivery date. Thus, in addition to monetary costs, time and convenience enter the equation for value also. The consumer's opinion of a product's quality could also change over time. A car which always provided a smooth, quiet rider can suddenly emit some screeching sound at high speeds, which cannot be satisfactorily fixed by the authorized service technician. Given, the various factors that come into play in formulating the concept of quality, it is not surprising that experts in the field have noted that while quality is "easy to visualize and yet exasperatingly difficult to define" [2].

A starting point for a discussion of quality must necessarily begin with an understanding of its different dimensions and how they are perceived. These dimensions are exposed through the various aspects of a consumer's interaction with a provider and its products or services. Scholars have identified five basic definitions for quality depending upon where one is viewing it from within the concept–creation–delivery–experience cycle.

The transcendent definition approaches quality from a philosophical perspective, an object's innate excellence. When a consumer purchases a product either for use or consumption, he/she is essentially buying a bundle of attributes. For a food item, the bundle can include nutrition, taste, aroma, etc. The bundle will be more complex for versatile products such as cell phones, which the average consumer uses for a multitude of reasons. The customer's first judgment of quality pertains to how well this bundle is put together. Here, the relevant attributes of the product are necessarily wide and all encompassing. For example, the benefits accruing to a consumer at a coffee shop pertain not only to the taste and temperature of the beverage which he/she was served, but also the cleanliness of the cup, the ability of the cardboard ring outside to insulate his fingers from heat, the availability of chairs where he could be seated, the view from the windows of the store, etc. The costs will include not only the price of the coffee, but also how long he/she had to wait, how efficiently the queue was organized, etc.

The product-based definition approaches quality from an economic perspective, quality is a precise and measurable variable. Goods and services can be ranked by measuring the aggregate quantity of anticipated attributes that they possess. The implication of this definition is that quality can be planned for and produced by a repeatable process. The beginning of this process starts with the visualization of the goods/services bundle one wishes to create. The bundle must incorporate the relevant dimensions of quality that are important to the consumer. Moreover, the design of the bundle should also ideally elicit a positive emotional response from consumers belonging to the target market. For products with physical attributes, this entails how the product looks and feels to the consumer, but also how it completes the tasks which the consumer expects of it.

However, the final product could be a service and have no physical attributes. These service offerings have different elements in the design map compared to physical products. For example, the service environment together with the provider's ability to be flexible in meeting special consumer requests is some of the important criteria to consider in service design.

The idea that the same level of quality can be repetitively produced leads us to a manufacturing-based definition: conformance to requirements. Once engineering has defined the design specifications for a product or service, a process can be designed to produce that product or service in a receptivity fashion. Any deviation from the established specifications implies a reduction in quality. While the association of quality with a product or brand starts at the design stage, the final delivery of quality to the consumer is affected by both the manufacturing and logistical process.

The user-based definition approaches quality from a perceptual perspective: quality is in the eye of the beholder. As individuals, we are all unique because of our genetic predispositions, cultural and social backgrounds, and our experiences. Because of this, each one of us can best be satisfied with a product or service that delivers value based upon one's own particular preferences.

The value-based definition views quality as a function of benefits received versus the costs of acquisition. Thus, a quality product is perceived as one that provides performance at an acceptable price, or conformance at an acceptable cost. The concept of value that is derived from the product or service is crucial in determining long-term consumer satisfaction with the brand name that epitomizes the product. The challenge for the company is in determining quality measurements that equate to value.

A traditional view of measuring quality has been the measurement of the final product or service after it has already been delivered to the customer. We emphasize the viewing of quality in a dynamic context. Thus, instead of simply concentrating on evaluating quality based on a sample of the output, the focus should be on continuously monitoring the quality of the product as it is being created. Thus, appropriate points must be chosen in the transformation process to accomplish this monitoring. Both the tangible and intangible elements of the total product must also be monitored.

The design of the product/service bundle and its transformation from raw materials to the final offering must be geared to the expectations of the target market. Providers must manage the external marketing of the product or service to ensure that the promise epitomized by the brand name and image is appropriate to what has been created. While the emphasis on physical products is necessarily on tangible elements, the final delivery of the planned service to the consumer has to be based on careful thought to the actual service encounter. Thus, other than the choice of the appropriate technology and associated equipment, service delivery has to also consider the selection, motivation, and monitoring of employees. These factors are related to the internal marketing of quality, where the provider ensures that all elements within the organization are capable of delivering the quality promised in external marketing campaigns.

In some cases, the various threads of quality and how they are intertwined might appear to be outwardly contradictory. For example, the engineering attributes behind the utilitarian component of product design might appear to be far removed from the artistic elements that are part of an attractive product appearance. However, both must be incorporated into a successful product design. These dimensions must be identified not only from the point of view of the consumer and the manufacturer, but also from the perspective of all stakeholders. These include not only the employees, but also the community and society at large. This book's premise is that quality can be successfully created only if all stakeholders have a stake in improving the final delivered quality. Also, all stakeholders must come to the realization that all parties stand to benefit from participating in the process of creating better quality products and services, marketed by the firm.

References

1. Reeves, C. A., & Bednar, D. A. (1994). Defining quality: Alternatives and implications. *Academy of Management Review, 19*, 419–445.
2. Garvin, D. A. (1984). Product quality: An important strategic weapon. *Business Horizons, 27*(3), 40–43.

Chapter 2
Creating a Competitive Advantage

Competitive advantage is the lifeblood of every company. Without a competitive advantage, it is next to impossible to attract and retain customers on a sustainable basis. Without a stable base of customers, the company cannot consistently earn profits. Without sufficient profits, the company cannot repair, or replace, assets as they wear out; it cannot develop new products and service offerings; it cannot hire and train new employees as the businesses grow or as employees turn over; it cannot re-engineer processes, devise new work methods, or purchase new technology as market demands change. In other words, without a competitive advantage to differentiate it from its competitors, the company will eventually go out of business.

One of the principle tenets of economics is that as a market moves toward a condition of perfect competition, profits will decrease to zero. This does not mean that the company is not making money, only that their profit margins are zero. With the ever-increasing pressures of globalization, companies struggle to maintain their competitive advantages. The reason for this is simple; humans are very clever, and once they see and understand something, they can duplicate it. In addition, with increasing levels of competition, the profits in a given market will decrease. Furthermore, with the increasing connectivity associated with the advanced telecommunication technologies that are driving globalization, the economic barriers to entry (e.g., time and distance) that traditionally protected markets, are no longer effective. Given these factors, companies must continuously develop new products and services that leverage their competitive strengths and reset the profit curve.

Any product or service provided by a company should be aimed at satisfying a set of basic needs or wants of customers. However, in most cases, the typical consumer has a large number of providers to choose from. Thus, it is imperative for the provider to be competitive in order to attract a sufficient number of customers. Retaining the loyalty of this group of consumers and expanding it will be instrumental in enabling the company to earn a consistent strand of profits. This

G.N. Kenyon and K.C. Sen, *The Perception of Quality*,
DOI 10.1007/978-1-4471-6627-6_2

stable bundle of financial resource should in turn be the wellspring from which the company can draw on to attract other resources, in order to stay ahead of the competition and be foremost in the mind of the target segment as the primary provider of the consumer's basic needs and wants.

The first step in gaining a competitive advantage is to understand the demand landscape. Here, the company has to understand not only the customer's basic needs they are striving to satisfy, but also the other important factors that could influence their choice. Next, the company has to survey the competitive offerings available for the consumer and ensure that its own offerings provide more value. While these steps appear to be relatively straightforward, many companies are often too centered on their own brands and totally misjudge both the demand landscape and the relevant competitive threats. Thus, it is pertinent to carefully evaluate the essential guidelines for creating a competitive edge for the company's products.

2.1 Understanding the Demand Landscape

Instead of focusing on its own brand offerings, the marketer must first understand the basic needs and wants of the target consumer segment that it aims to serve. The two principle elements associated with the design of an offering are as follows: what is the primary need of the customers and what are the key differentiators for this market segment? For example, a restaurant's primary purpose is to eliminate the customer's task of preparing and cleaning up after preparation of a meal; the specific cuisine offered by the restaurant is a means of differentiating itself from its competitors. While this might seem obvious, the principle should apply to all product and service offerings.

In another instance, airlines exist to satisfy the need for large-scale, fast, and convenient transport from one part of the world to another. Differentiators in this service range from price, to scheduling, number of destinations, and in-flight services and comfort. Thus, when it started its services, Southwest Airlines (SWA) adjusted an important factor related to their target market's travel plans between Dallas and Houston. In addition to offering their "peanut fares" (i.e., low price), they also changed the airports their consumers used in their itinerary. Instead of driving longer distances to airports to catch their flights, travelers now got the option to go to airports such as Hobby and Love Field which were situated at more convenient locations. So, rather than concentrating solely on actual flying time between the two cities of Dallas and Houston, the management of SWA looked at the broader issue of the total travel experience between these two destinations.

When a company focuses on the basic needs and wants of its target market, it must strive to consider all factors that enter into the interaction between the product and the customer. In many cases, some of these factors might not appear to be related to the basic needs and wants. For example, a restaurant serving food at its current prices could see its sales volume drop because of increased crime in the neighborhood. In order to maintain a healthy, stress-free dining experience, the restaurant management might consider providing more security near the restaurant's dining and parking areas or provide valet parking services. In extreme situations, it could consider changing its location. In this case, the impact of location increased because of circumstances beyond the restaurant management's control. While some attributes such as the quality of food and service for a restaurant are expected to remain essential to the customer over time, other attributes such as the provision of additional parking spots or extra security might grow in importance.

In some instances, as in the case of smartphones, new features such as the ability to text and the availability of multiple applications (apps) can grow to be as important as the original basic quality of making and receiving phone calls. Thus, all products or services need to be periodically reviewed to insure that current elements of their design still play a part in consumer choice. This understanding of the demand landscape is the first step in the attempt to gain a competitive advantage. Here, the important criterion is the relative importance of each attribute and their interaction with each other. As described in the examples cited above, the relative importance of every attribute will vary with time. However, in mapping this network of needs and product charteristics and sevice attributes, the company must be cognizant of what the customer's basic needs are and how their product or sevice is expected to meet those needs.

2.2 Choosing the Competitive Space

When analyzing the demand landscape, it is important to note that consumer perceptions of how each product lives up to its expectations can be quite different from what the manufacturer thinks is the case. Thus, as a first step, a customer perception map of each product/service offering must be drawn. Here, it might be necessary to draw different perceptual maps for the different target markets that might exist for the product. In addition to using the demographic factors, such as age, income, gender, and ethnicity, the marketer can use its own intuition to choose the target markets. The perceptual maps should be based on analytical techniques such as factor analysis and/or multidimensional scaling.

Based on the results of the perceptual maps, the marketer can better understand both who its nearest competitors are and gain insights into potential new products that can occupy existing empty spaces in the perceptual maps for various target markets. The next step for the marketer is to aim to deliver the product that excels in providing value to the chosen target market. In order to do so, the marketer must carry out a conjoint analysis. An important initial step in carrying out this analysis is to choose the relevant product attributes and provide the consumer group evaluating the alternatives with a comprehensive but concise set of theoretical product alternatives. The respondents will then rank these different alternatives from the most to the least desired. In most cases, product price is one of the alternatives. It is important that the marketer has realistic estimates of the price points of each product alternative. The conjoint analysis procedure will provide "utility" values for each attribute included in the product alternatives. These values are essentially the output from the different trade-offs made by the respondent group in their ranking of the product alternatives. Based on the chosen utility levels for each attribute, including price, the marketer can decide on the combination of attributes together with the desired level of excellence required on each of them as a harbinger of what their product offering must aspire to be. The marketer must ensure that the excellence level at the desired utility points for each attribute are met by every item in their product line offered to their target market, thus, establishing the initial quality level for the products and/or services in the analysis. However, consumer evaluations of the balance between the benefits of the network of attributes and the various costs associated with them are likely to change over time. Because of this time phased variance, the quality level for each attribute can never be permanent. The marketer must constantly evaluate consumer evaluations to understand the quality level on each attribute it should aim for. Thus, quality aspirations for every product are never static but always dynamic in nature, as a response to changes in the environment, particularly because of technological and competitive shifts.

2.3 Defining Competitive Advantage

For decades, it has been espoused that companies are individually unique because of the mix of people and processes that constitute the organization, with people being the primary source of a company's competitive advantage. Though this is true, the products and services that a company markets to its customers are produced and delivered through the execution of processes, and processes are not unique. Unfortunately, both processes and their outputs (e.g., products and services) can be duplicated, and it is the products and services that are typically the basis of the customer's quality assessment of the company's brand.

If the product, or service, design meets customer's requirements and expectations, then properly designed transformational and delivery processes will insure that the intended functional benefits of the product and/or service are delivered. The question now is whether or not the customer perceives the benefits as meeting his/her expectations. If they do, then the customer will perceive the product and/or service as being of good quality and will be satisfied; if no, then the customer will be dissatisfied.

Regardless of the originality or innovativeness of the purpose and regardless of the fact that there may be numerous methods of producing the same results, there is only one optimal method for meeting a given objective. This optimal method will produce the desired results more efficiently than any other methods, thus yielding a competitive advantage with respect to cost and functionality.

In striving to achieve a competitive advantage, companies must recognize that customers do not buy products or services, they buy value, and value is a function of quality and price. The pricing of a product or service is driven by the cost of production, the size of the market, competitive actions, and other factors. Quality though is more difficult to quantify. Not only must product and service designers understand customer requirements and usages for the product or service, but they must also understand how the functionality of the product or the benefits of the service are perceived by their customers.

In the strategy literature, a common theme associated with competitive advantage is value creation. Competitive advantage has been described as the basis for the firm's performance in competitive markets. It is argued that competitive advantage is created and sustained through the value the firm creates for its customers in excess of its costs of creating said value. From a managerial perspective, a competitive advantage is the ability gained through attributes and resources in achieving a sustainable profit margin greater than the average profit margin of competitors within the same market. From the customer's perspective, a company has a competitive advantage when it has a sustained ability to create value for the customer at a rate greater than its competition. The major problem with understanding competitive advantage is in understanding value and how to create it. So, who is the arbiter of value? The simple answer is the customer.

2.4 Creating Competitive Advantage

The best strategy for the creation of competitive advantage is to deliver quality products and services at a competitive price. But, is this enough? No, you must deliver products and services that the customer perceives as higher quality than the alternatives. These products and services must also be perceived as being capable of being beneficial in satisfying their needs. One would think that if the product or service is actually of higher quality than competitive product, then everyone would perceive that fact. Unfortunately, because of our diverse worldviews and motivation, that is not true. To assume that quality is recognized the same by everyone violates the first law of reality.

Reality is not What Actual Occurs; Instead, It is What is Perceived to Have Occurred

Take the following example: Today everyone knows that four-wheel brakes are superior to two-wheel braking systems. This fact has always been true but has not always been perceived as being true. Take, for example, Rickenbacker Motor Company. This company used innovation as their competitive advantage, adopting it into both their designs and their engineering. Most of these innovative changes were readily accepted by the market. But in 1924, Rickenbacker equipped their new line of cars with four-wheel braking systems. This innovation was a major change from the common practice of vehicles only having a rear wheel braking system. This change from current practice allowed competitors to argue that four-wheel brakes were dangerous since they might put the car into a skid or, if they worked as designed, would throw passengers into the dashboard. Because of these criticisms of this innovation, Rickenbacker sales in the mid-1920s slowed, placing the company in financial troubles. Due to similar management miscues, the Rickenbacker Motor Company eventually went out of business. The year following the discontinuation of the four-wheel braking system by Rickenbacker, the firms that so ardently denounced that system incorporated it into their new car models and were able to successfully market the product. If Rickenbacker had factored consumer perceptions into the equation and worked to educate the buying public of the advantages of this safety feature, Rickenbacker Automotives might still be a major player in the automotive industry today.

2.4.1 Competitive Strategies

Strategy is focused upon understanding what customers want and systemically developing plans that align organizational resources and policies to deliver it to them. Companies typically have multiple layers of planning that are differentiated by both their time horizons and the breath of their objectives. Strategic planning has the longest time horizon, and its primary objective is the securing of resources necessary for the future creation and maintenance of a competitive advantage and the setting of priorities with respect to the usage of those resources.

The classical approach to developing competitive strategies is the structure–conduct–performance (SCP) paradigm. There are three principle elements to this paradigm: the structure of the industry, the conduct of the firms in that industry, and the performance of the firms within the industry. This paradigm suggests that structural changes will impact the conduct of the firm and the firm's performance and that changes in the firm's conduct will impact performance. A major criticism of the SCP paradigm is that "in practice, the firm's actions (conduct) and profitability (performance) can influence market structure."

The main assumption of the SCP paradigm is that market power is directly (positively) related to profitability. The causal relationship assumed in the SCP paradigm typically holds in homogeneous industries, rather than heterogeneous industries. A fundamental assumption of the SCP paradigm is that the most important factors in defining an industry's structure are the barriers to new entrants. The principle barriers to entry defined by the SCP paradigm are as follows: economies of scale, product differentiation advantages, and absolute cost advantages.

There are fourteen factors that are the common causes of entry barriers. Exogenous (economic or intrinsic) causes include capital requirements, economies of scale, product differentiation, absolute cost advantages, diversification, research and development intensity, high durability of firm-specific capital, and vertical integration. Endogenous (voluntary and strategic) causes include retaliation and preemptive actions, excess capacity, selling expenses (including advertising), patents, control over other strategic resources, and the scope of product offerings.

There are three basic strategies recommended for generating competitive strategies: the lowest cost, differentiation, and focus. These generic strategies are defined along two dimensions: scope and strength. Strategic scope is a demand-side dimension that captures the size and composition of the target market. Strategic strength is a supply-side dimension that defines the firm's core competencies. In formulating how to best apply a given strategy, Porter pointed out that there were optimal markets to strategy relationships. If a firm wished to serve a broad market with a low-cost strategy, it would need to establish a cost leadership position. On the other hand, if that company wanted to use a differentiation strategy, it would need to focus on how to differentiate its products and/or services from the competition. If the firm's chosen markets were narrow in scope, then a similar strategic approach would need to be tailored to each of those markets.

The differentiation strategy involves creating products that present customers with a higher-value proposition than other products in the market. This strategy typically works best in markets where the customer is not price sensitive, or the market is competitive or saturated, or where customers have specific requirements and the firm has unique resources and/or capabilities which enable it to satisfy these requirements in ways that are difficult to copy. A dressmaker could create innovative designs because of its studio of creative artists who constantly turned out dresses which caught the public's attention. The particular studio could attract creative designers because of their extrinsic and intrinsic rewards. Creative designers considered it a boon to their careers to be working for this particular studio.

Within the focus strategy, there are three variants: a cost focus, a differentiation focus, and the cost and differentiation focus. This strategy focuses on describing the scope over which the company should compete based on cost leadership or differentiation. By targeting narrow markets (also called a segmentation or niche markets), the company could choose to compete on the basis of either low prices or differentiated depending on the needs of the selected market and the firm's resources and capabilities. Although there were few concert pianists, a software company could be successful by creating software that catered only to their interests, by making products that, through sound recognition, removed the chore of turning the pages of sheet music during rehearsal.

The implementation of total quality (TQ) directly influences the firm's competitive position, thereby constraining the strategic options available to the firm. TQ can effectively govern much of what conventionally required executive-level strategic planning. The implications of this comment are that if the company is continuously improving its quality, other strategic considerations may become unimportant. Basically, with TQ properly implemented, the firm will have a corporate-wide focus on the customer, will implement quality starting at the initial product/service planning, and will produce the product/service right the first time. Thus, the firm will be in the position of applying simultaneously a cost leadership strategy and a product differentiation strategy. As noted in the Rickenbacker automotive example, "higher quality does not ensure competitive success; marketing issues such as timing and technical standards can undermine even the finest of products". Therefore, the implementation of strategy requires a holistic plan addressing the customer's perceptions as well as market conditions and timing issues.

2.4.2 The Competitive Dilemma

Economic theory tells us that as competition in a given market increases, the profits in that market will decrease. This occurs because as competitive pressures increase, companies will decrease pricings, striving to generate more sells. Eventually, this behavior, in the long term, results in zero profits. To negate this result, companies must continuously introduce new products and services, thus resetting the price curve. The dilemma comes in with customer expectations. Each time a company introduces a new product or service, it must re-excite the customer, but by doing so, the company resets the bar for what the customer will expect the next time. If the features of the new product or benefits of the new service are not sufficiently different, and better, than the previous model, the customer will perceive the deceit and will abandon the company. If the features of the new product or benefits of the new service are sufficiently different, and better, than the previous model, the customer will expect even more with the next product/service offering. Furthermore, if the industry segment as a whole introduces to many products or services into short a time period, the customer will start suffering fatigue.

Chapter 3
The Value Proposition

As mentioned in the introduction, the purpose of any business is to generate money. The logic behind this objective is that without financial viability, the business cannot maintain or grow resources. The business will not be able to develop new products or services, hire, and pay employees or any other activity that society regards. The question is how does a business consistently generate money? The simple answer is through the creation of value for its customers, employees, investors, and the country. Furthermore, due to the interests of these three groups being inextricably linked, sustained value can only occur when value is created for all three groups simultaneously.

So, what is meant by value? Historically, value has been defined along two concepts: usage value and exchange value. Usage value is a perceived assessment that is based upon the perceived benefits of a product or service. This measure of value is subjective and changes over time. Exchange value is a quantitative assessment based upon the revenue received from the sale of a product or service in the marketplace. This value can only be generated once per unit. These traditional views of value are very limiting to a company interested in improving its competitive positioning and improving it customer's satisfaction. Ultimately, customer value is conceptual in nature and is often thought of as the different between the perceived benefits of a transaction and the sacrifices associated with obtaining those benefits.

3.1 Benefits and Sacrifices

The benefits received by the customer that are inherent to the product are performance, conformance to specifications, reliability, and safety. In contrast, the perceived strategic benefits for the provider are its expertise, competencies, the advantages its products have over its competitors, and the availability of new products.

Co-Authored with: M. Oduwole, M.S. I/O Psych., Lamar University.

© Springer-Verlag London 2015
G.N. Kenyon and K.C. Sen, *The Perception of Quality*,
DOI 10.1007/978-1-4471-6627-6_3

The perceived personal benefits to the customer are pleasant experience, satisfying, personal value, and recognition. The sacrifices made by the customer are the time and effort expended on the product search, the monetary price paid for the product or service, and an opportunity cost associated with the loss of functional benefits should the product or service be less than perfect.

3.2 Value and Motivation

Value is defined as "desirable, transituational goals, varying in importance, that serve as guiding principles in people's lives." The main factor in understanding value and the differences within these value types is to understand the motivational goals that they are an expression for. At its basic level, human existence centers on biological needs, social interaction, and survival. There are ten types of motivational factors: power, achievement, hedonism, stimulation, self-direction, universalism, benevolence, tradition, conformity, and security.

The strength of these values in motivating individual actions is strongly related to their cultural upbringing. In horizontal societies as in many Asian cultures, individuals value equality and tend to view themselves as having the same status as other members within that society. Individuals that are more concerned with welfare than with their own advancement are motivated by self-transcendent values. In vertical societies as in Western cultures, individuals see themselves as different from other members of that society and accept inequality and believe that rank has its privileges. Individuals that are more concerned with their own welfare than with their own advancement are motivated by self-transcendent values; whereas individuals that are concerned with their own welfare and success are motivated by self-enhancement values.

3.2.1 Motivational Theories

Motivation is what drives people and why people think and behave as they do. In the literature on motivation, numerous theories have been developed. Three of the more noted general theories on motivation are: Clark Hull's drive theory, Kurt Lewin's field theory, and Victor Vroom's expectancy theory.

3.2.1.1 The Drive Theory

Whenever a person is stimulated by a desire and by a specific goal and that goal is achieved by a given response, the bond between the stimulus and the response is strengthened. When the goal is not achieved, the bond is weakened. Hull asserts that for a prior association to be displayed, there had to be some unsatisfied need driving the action. This drive seeks to return the individual's physiological state back to equilibrium. Mathematically, this behavior is modeled as follows:

$$\text{Behavior} = \text{Drive} \times \text{Habit}$$

For example, the needs of a very young baby are few and unlearned. From time to time, he needs milk (drive). The milk is supplied by the mother, who usually appears just before the milk (habit). After a number of such experiences, the child learns to want the mother herself and her presence becomes satisfying even when the baby is not hungry (behavior).

3.2.1.2 The Field Theory

Kurt Lewin's field theory assumes that behavior is determined by the totality of an individual's circumstances. In this theory, 'field' is defined as 'the totality of coexisting facts which are conceived of as mutually interdependent.' Lewin believed that individuals behaved in accordance with the way in which their tensions (the magnitude of need) between their perceptions of self and of the environment were internally processed. Mathematically, behavior is modeled as follows:

$$\text{Behavior} = F(\text{Person, Environment})$$

Therefore, the motivational force to act is determined by the magnitude of the individual's need (tension), their goal (valence), and the perceived psychological distance between themselves and their goal which are modeled as follows:

$$\text{Motivational Force (MF)} = F(\text{Tension/Valance})/e$$

The logic driving this theory is that when a person needs or desires something, he or she is in a state of tension. A need is Lewin's basic motivational concept. It may arise from a physiological condition such as hunger or may be a desire or intention to do something. Needs release energy, increase tension, and determine the strength of vectors and valences.

The field of the person is in a state of tension whenever a need exists. A positive or negative valence is the attraction or repulsion that a region in the psychological environment has for the individual. Positive valence exists when the person believes that the region will reduce tension by meeting their needs. Negative valence occurs when the person believes that the region will increase tension or threatens injury.

A vector is a force that arises from a need that acts on the person and determines the direction in which he or she moves through the psychological environment. For every region with a positive valence, a vector pushes the person in its direction. With a negative valence, a vector pushes the person away from it.

We build tension in order to motivate ourselves to learn and experience new things. When we complete tasks, the tension is released. This sense of relief is related with the closure that is acquired when you finish what you start. Life is a constant interchange between completing old situations and introducing new ones. If we are alive and well, then there is always excitement, tension, and possibilities.

You can get closure and reduce tension, but the tension is never eliminated because we keep our systems open to be able to explore new events, people, and possibilities. This theory also applies to cases where a person is focused on completing a task or solving a problem. An example here is a teenager who would like to see his/her favorite music group when they are touring his/her hometown. The tension rises when the teenager hears that the tickets are all sold out. Thus, when a friend gives him his ticket because he was going to be out of town, the teenager feels relieved and happy. However, as the ticket is in the "nose bleed" section, the teenager still feels inclined to see if any of his other friends and acquaintances will give him an opportunity to move to a better seat within the theater.

3.2.1.3 The Expectancy Theory

Proposed by Victor Vroom in 1964, the expectancy theory states that the intensity of an inclination to perform a task is dependent on the intensity of the expectation that the successful execution of the task will result in a certain outcome and on the strength of the appeal that the outcome has for the individual. In other words, a person's motivation for producing a given outcome will depend on the desirability of the reward offered (valence), the likelihood that the effort will result in the expected outcome (expectancy), and the strength of their belief that achieving the outcome will actually lead to receiving the reward (instrument). Mathematically, the expectancy theory can be modeled as follows:

$$\text{Motivational Force (MF)} = \text{Valance} \times \text{Expectancy} \times \text{Instrumentality}$$

The expectancy theory concentrates on three relationships:

- The effort–performance relationship: This relationship focuses on the likelihood that the individual's effort will be recognized in the performance appraisal. Basically, there is a high probability that the individual will be able to perform the behavior if they try.
- The performance–reward relationship: This relationship focuses on the extent to which the individual believes that getting a good performance appraisal leads to the rewards.
- The rewards–personal goals relationship: This relationship focuses on the attractiveness of the potential reward to the individual.

If any of these three relationships are lacking, the individual is unlikely to direct his or her efforts toward the particular action. For example, an employer might say that productivity is down because of long lunch breaks. The employer is willing to give employees a 5 % raise to shorten their lunch breaks. Valence is measured from -100 to 100. In this instance, valence could be 50 (how attractive the reward is). Expectancy is measured on a 0–1 scale. Expectancy here could be 0.9, the likelihood that the individual would satisfy the criteria to receive the reward. Instrumentality is also measured on a 0–1 scale. Instrumentality is the probability that the employer will do what they promise, which could be 0.8 in this situation. To calculate the motivational force for employees to shorten their lunch break, you would have (50) (0.9) (0.8) = 36. Motivational forces max out at 100, so a force of 36 will most likely not encourage employees to direct their efforts towards reducing the time spent at lunch.

3.2.2 Types of Motivational Goals

Research has found ten common value motivators that are consistent between individuals on the things they value in life. These stated values, or motivators, were also found to be related to each other and are reoccuring across time, countries, and cultures. These ten value motivators are described as follows:

The first type of value motivator is power. It is based on social status and prestige, and control. As a concept, power is value dependent. There are several types of power: confirmed power, wasted power, hidden power, and unknown power. Confirmed power is recognized by, both, you and others. Wasted power occurs when others recognize that you have it, but you do not. Hidden power occurs when you believe you have it but others do not. Unknown power exists when you and others do not believe you have it.

There are six stages in the development of personal power: powerlessness, association, achievement, reflection, purpose, and wisdom. The individual's initial stage is powerlessness. The individual will remain in this stage until he learns how things work in his environment and gets to know the other members in his environment and how to influence them. The next stage of power development comes from those we associate with. By gaining the respect and trust of those in our environment, the individual can start leveraging their power. The third stage of power is achievement. This power is acquired through our actions. Our successes become persuasive evidence that others will perceive as leading to more success. This type of power will multiply as others cede their power to those who prove their ability, which in turn allows them to achieve even further. The fourth stage of power is reflection. By reflecting on what we have seen, done, and learned, we realize that we have a personal power upon which we can draw. Individuals at this stage of development are often competent and display sound judgment and integrity. The focus and commitment associated with striving for a higher purpose, or struggling for a just cause, create power. Because this inner power is often so much greater than the power of others, it can influence their decisions. Examples of this type of power can be seen in great leaders when they show purpose in a stirring speech or in powerful and symbolic actions. The sixth stage of power is wisdom and comes when people feel a deep connection with the universe or some spiritual source. These individuals often display contentment and live life on an 'even keel.' They are capable of knowing and accepting powerlessness and in doing so frequently find ultimate power.

The second type of value motivator is achievement. It is focused on the need to demonstrate one's competence as measured by some social standard. Achievement is positively related to the personal value of self-respect, as well as the cultural indicator of economic development. The values of achievement and power are positively related to vertical cultures, but are negatively related to horizontal cultures. The goal of the achievement value type is personal success, and this goal is reached through the demonstration of competence as defined by some social standard.

The third type of value motivator is hedonism, which is centered on self-gratification and pleasure. It has even been argued that pleasure is the only intrinsic good. This school of thought states that people are motivated by the need to maximize value for themselves and thus find the maximal balance of pleasure over pain.

The fourth type of value motivator is stimulation. Though similar to hedonism, it finds value in excitement, novelty, and challenge. Presumably, variety and stimulation help us maintain optimal levels of activation and lead us to possess stimulation values which are likely related to self-direction values. Our need for control and mastery, as well as autonomy and independence, lead us to desire self-direction, which involves creativity, exploration, freedom and independence.

The fifth type of value motivator is universalism, which values understanding, appreciation, and tolerance through wisdom and an open mindedness, as well as protection for the welfare of all people and for nature. People recognize that neglecting nature will lead to loss of natural resources on which life depends. By the same token, having intolerance for individuals outside of one's immediate group will lead to life-threatening strife. These values are virtually present in most cultures, with the exception of isolated, homogenous, smaller ones. Universalism involves the maturity value type which seems to emerge from human requirements empirically (verifiable through observations) rather than a priori (theoretical deduction). Also included is a part of prosocial value type. When groups or individuals recognize scarcity of natural resources after coming in contact with other individuals who are outside of their immediate group, they experience survival needs. Those survivals needs give rise to the motivational goal of universalism values (i.e., tolerance, protection for the welfare of all people and for nature).

Research stresses importance on differentiating between the two types of prosocial concern (universalism and benevolence) based on collectivist versus individualist types of cultures. While the welfare of members belonging to the same cultural group is taken into great consideration in collectivist societies, relative needs of members of the out-group are treated indifferently. However, this is not true in individualist cultures where less emphasis is put on the distinction of in-groups' and out-groups' needs. According to this pattern, collectivist cultures focus more on benevolence than on universalism values, and individualist cultures do not weigh either value type as higher than the other.

There are eleven types of motivational values that address three basic theoretical questions: (1) Are all of the value types present as distinctive organizing principles in all samples? (2) Do the same specific values constitute each motivational type in each culture? and (3) Are any other value types necessary to account for the organization of single values? Within each sample, correlations among single values are represented by multidimensional space. By examining two-dimensional projections of the multidimensional space, one can operationally derive evidence for or against the existence of value types, as well as their consistency of single values. It is possible that there will be variation in the location of single values and the distinctiveness of types. In situations like this, shared organizing principles at a greater abstract level are suggested, as well as broader categories that exhibit more universality.

The sixth type of value motivator is benevolence. Benevolence seeks to preserve and enhance the welfare of others. More specifically, it focuses on concern for welfare of individuals close to oneself, with whom one interacts on regular basis. This concept, thus, makes benevolence a more focused and narrowly defined version of the prosocial value type, which focuses on well-being of all individuals, no matter the setting or situation. Since benevolence is more tightly focused, its motivational goal values of helpfulness, loyalty forgiveness, honesty, responsibility, true friendships, and mature love support Schwartz and Bilsky's need of promoting and flourishing of groups, as well as need for affiliation.

The seventh type of value motivator is tradition. This motivator is driven by respect, commitment, and acceptance of the customs and ideas of a culture or religion. Religious rites, beliefs, and norms of behavior represent a group's unique worth and solidarity and are the most common forms of tradition. Symbols and practices are developed by groups, which will serve to represent its tradition and customs and will become valued by its members.

The eight type of value motivator is conformity, which involves self-restraint of ones actions, inclinations, and impulses. It engenders politeness and obedience to socially recognized elders and aims to avoid violation of social expectations and norms. By practicing conformity, one accepts others' behavior as his or her own in order for things to run smoothly within a group or a setting. This is the ultimate and defining goal of conformity. The underlying basis here is that the individual assumes that other members of the group know something that he or she does not.

Security is the ninth type of value motivator, and it is defined by safety, harmony, and personal relational and social stability and thus has the same motivational goals. Each of these factors will stimulate a perception of value in and of themselves. Higher levels of value are achieved through the dynamic coupling of two or more of these factors. For example, combining achievement and power motivators can lead to an emphasis on social superiority and esteem; achievement and hedonism can lead to narcissism; stimulation and self-direction are intrinsic motivators of mastery and openness to change. Similarly, negative results (as seen by society) can be experienced through the combining of factors.

The tenth value motivator is self-direction. With the motivator the individual seeks independence of thought and action. The theory of Self-Determination asserts that humans have persistent positive features, which repeatedly show effort, agency, and commitment in their lives. People also have innate psychological needs that are the basis for self-motivation and personality integration. The basic needs are competence, relatedness, and autonomy. These needs must be satisfied in order to foster well-being and health. The need for competence seeks to control outcomes and experience mastery. The need for relatedness wants to interact, be connected to, and experience caring for others. While the need for autonomy is the urge to be the causal agent of one's own life and to act in harmony with one's integrated self.

3.2.3 The Dimensions of Value

Our overall assessment of product's utility is based upon our perceptions of what is received and what is given. We term this assessment as value. The concept of value has traditionally been defined along two dimensions: economics and psychological. Even though the concept of value has evolved beyond its original definition, it is still inextricably linked to the constructs of quality and satisfaction.

Personal values have been found to be significantly associated with product attributes and expectations in industries ranging from automotives to food, to travel decisions, to media preferences, and to the usage of mass media. Furthermore, the relationships between value to consumer beliefs, preferences, and positions on social issues were significantly different. It has also been found that the customer's value constructs explain their consumption behaviors. These behaviors range from product choices, purchase intentions, and repeat purchasing.

From a business perspective, customer values can be used to segment markets. Personal values and expectations about the benefits of a transaction play a significant role in consumer behavior. In order to design and manage products and services that drive customer satisfaction and loyalty, it is accentual to understand the dimensions along with customers measure value and the core psychographic variables that make up these dimensions.

Our understanding of value has evolved from its two-dimensional roots to a hedonic versus utilitarist value dichotomy. This dichotomy considers efficiency, excellence, play, aesthetics, esteem, status, ethics, and spirituality using a three-dimensional paradigm: intrinsic, extrinsic, and systemic. The intrinsic dimension of value encompasses the realm of uniqueness and singularity. This dimension perceives value in persons or objects based upon singularity, essence, uniqueness, or spirituality.

The extrinsic dimension of value is the domain of comparisons. In this realm, measurements of value are rooted in good, better, and best assessments. The extrinsic dimension is one of results and common sense where tactical planning, role satisfaction, and social fulfillment exist.

The systemic dimension of value focuses on formal concepts, such as definitions, ideas, goals, structured thinking, policies, procedures, rules, laws, and should and should not's. This is the dimension of perfection.

These three basic dimensions of value are applied both internally and externally, thus creating six core dimensions: (i) intrinsic–external, (ii) intrinsic–internal, (iii) extrinsic–external, (iv) extrinsic–internal, (v) systemic–external, and (vi) systemic–internal. All six of these core dimensions are utilized any of the thousands of subconscious daily decisions. What distinguishes each of us in this process is the ratio with which we apply these dimensions.

Research has found that price perceptions are highly context specific. In fact, time and effort will affect the perception of cost in a transaction. The perception of a product or service's quality is an antecedent to the assessment of its value, and satisfaction is a consequence of this value assessment. Loyalty is often the result

Table 3.1 Attributes of value

| Dimensions of value | | | | | |
| Intrinsic (People oriented) | | Extrinsic (Task oriented) | | Systems (Environmental) | |
External source of value	Internal source of value	External source of value	Internal source of value	External source of value	Internal source of value
Empathy	Accomplishment	Rationality	Belonging	Ethics	Goal
Fun/ enjoyment	Cooperation	Recognition	Respect	Justice	Curiosity
Excitement	Happiness	Reputation	Relationships		Independence
Security	Self-esteem	Materialism	Convenience		Creativity
	Self-fulfillment	Success	Quality		
	Spirituality	Power			
	Trust	Authority			

of sustained satisfaction. When relating satisfaction and loyalty to the dimensions and attributes of value (Table 3.1), extrinsic attributes such as efficiency and quality have been found to be related to loyalty behaviors and intrinsic attributes such as social value and play are related to satisfaction.

The quality assessment of products and services is often multidimensional in nature and in many cases is not constant over time or equivalent across demographically diverse groups. Understanding the variations in individual perceptions and product (service) characteristics (attributes) can provide important insights for design and/or improvement activities as well as for management and marketing.

3.2.4 Types of Value

There are several types of value. Financial value is composed of revenues that the company receives from its products and services in the marketplace. There is strategic value which is associated with the company's ability to acquire and maintain a technology base for a broader industry and the extent to which the company is investing in long-term research and development. There is also social value which is the degree to which the company engages in activities that society values. In order to be viewed as a high-value organization, the company must maximize value for its stakeholders. Figure 3.1 presents a matrix between the four principle stakeholder groups and the three value types. In assessing the degree to which a given action on the part of the firm creates value for its stakeholders, it needs to map the resulting benefits of the action back to each group, making sure that the benefits each group recieves are consistent with their respective needs.

How a company generates value will depend upon which stakeholder it is focusing on. Each class of stakeholder will value different things and has different expectations of value. Internal stakeholders will see value with improvements in

		Value Types		
		Financial	Strategic	Social
External Stakeholders	Customers			
	Society			
Internal Stakeholders	Investors			
	Employees			

Fig. 3.1 Basic value matrix

the organization, its processes, and its financial viability. Employees receive value from being treated with respect, being involved in decision making, engaging in meaningful work, and from continued training and other developmental activities. Other areas where employees receive value are as follows:

- Competitive salaries
- Retirement programs
- Safe work environments
- Professional growth opportunities
- Personal growth opportunities
- Social interaction and team-based activities.

Investors realized value when the company delivers consistent high returns on their investments. They also receive value from the following:

- Risk-adjusted financial returns
- Ethically business practices
- Long-term growth
- Reinvestment
- Strategies and practices that improve adaptability and sustainability
- Promotion of employee loyalty and ownership
- Transparency in corporate governance
- Timely communications.

External stakeholders receive value from the benefits derived from the company's externally facing activities, its products, and its services. Customers receive value when the company develops and markets products and services that are consistently useful. The country and society in general receive value when the company provides:

- Sustainable employment
- Pays taxes
- Creates a positive trade balance
- Positively impacts the gross domestic product (GDP)
- Environmental leadership
- Develops new technologies and other forms of intellectual capital
- Reinvests in science and engineering
- Inspires the next generation.

In using the matrix, strategists should identify performance metrics that measure where the company is currently for as many value creation activities (e.g., relevant to active and future markets the company operates in) as possible for each stakeholder/value-type intersection. Then, goals should be set for future performance. To best understand how well the company is progressing to its goals, a unidimensional metric can be calculated as follows:

- Verify that all measurements are positively correlated. For measurement that negatively correlated, modify by multiplying measurement by -1.
- Rank all measures, variable by variable, on a 0–100 percent scales.
- Use either weighted scores or averaging of the resulting % ranks to obtain the composite score.
- Using the resulting composite measure for the analysis. In case of ties, we assigned the maximum, so that top performers are assigned 100 %.

One of the biggest impediments to a company's value-creating process is the failure of top management to define the mission of the company properly. Instead of narrowly defining the company's mission as the maximization of shareholder wealth and setting short-term financial goals to accomplish the mission, the mission should be stated in terms of value creation and the strategies for achieving it should be defined in terms of value-adding activities. When the firm systemically creates value for its customers and employees, the financial returns will be realized, creating value for shareholders.

Managers will often choose not to focus on value creation for two principle reasons. First, their training and education defined business success too narrowly. Understanding the needs of each stakeholder group while engaging in continuous innovation and creation of new of new products and services, and the designing and managing processes is hard work. Furthermore, the right answers are frequently not the obvious answers, and solutions usually take a long time to achieve. Secondly, the financial markets and conventional industry economics are focused on short-term thinking and financial results. This behavior is also supported by our financial accounting systems not to mention that financial measures are quickly obtained and easily understood.

As a result, they systemically decrease the long-term value of the firm. The long-term solution is for companies to define their interest more broadly. Overall, mission statements and corporate strategies need to include the interests of customers and employees. In his discussion of this concept, Paul O'Malley notes that, "highly motivated, well trained, properly rewarded employees deliver outstanding service, while effective R&D investments lead to products that enjoy a significant value-adding advantage and generate higher margins. Satisfied, loyal customers drive revenue growth and profitability for investors." Clearly, a win-win for everyone: except for the competition.

An important factor in the decision-making process is the time frame of the decision, because it has a significant influence on one perception of self-interest. If you knew that 95 % of increased volume over the time period covered by running a volume discount promotion would be from a shift in when customers made their purchases instead of from new customers, would you rub the promotional? If you knew that running the same promotion multiple times would generate the same results, would you run the promotional? If you knew that running the promotion routinely on the same frequency, with the same results, would you run the promotion? Most likely, you would say yes to each of these questions, and your reasoning would most likely be, "because 5 % of the sales volume was from new customers."

Now, consider the demand planning process at Compaq Computers, before its merger with Hewlett Packard. Somewhere in the distant past, Compaq needed to generate more sales in order to meet its end of quarter revenues so as not to negatively influence stock prices. To do this, they ran a volume discount promotion for their channel partners. The increased sales meet the revenue requirements, and the company's stock was rewarded with an increase in price. Unfortunately, the increased sales in that quarter came at the expense of sales in the next quarter. Sales in the first month of the following quarter were down from forecasts; no big deal, sales should pick up in the second month. The second month sales were good, but not good enough to get the revenues for that quarter back on track; so, management runs another promotion, sales increased, revenues meet predictions, and the shock price increased. The down side of these decisions was that the customers saw the pattern and started waiting to place their sales orders till the end of the quarter in anticipation of Compaq running another volume discount promotion. Guess what, management blinked, ran the promotion, rewarding the astuteness of their customers.

Three things occurred here and neither where good for the Compaq Company. First, by routinely running the volume discounts in order to save the short-term stock price, management artificially changed customer buying habits. Secondly, this change in buying caused greater fluctuations in forecasts, thus creating greater uncertainty in demand and creating non-optimal operational planning, which added expense to the overall operations and negatively impacted employee morale and productivity. Finally, the changes in customer buying patterns caused an overall decrease in revenues per unit volume because of the discounting. The customers won, the employees lost, and the investors at best came out neutral. So, would you still say yes to the previous questions?

The key element to take away from this is that management should focus more on the creation of opportunities than on minimizing costs or maximizing short-term gains. If Compaq's management had stayed focused on long-term value-adding strategies and managed the company's day-to-day business toward achieving those long-term, value-based strategies, the short-term results would have taken care of themselves. The major themes associated with value creation are as follows: product and process innovation, monitoring of changing customer needs, leveraging of emerging technologies in existing markets, leveraging of technology and regulatory changes to create new markets, re-engineering and managing the company's value chains, and creating win/win partnerships with customers, employees, and suppliers.

3.3   Value-Creating Strategies

Customers are no longer just passive consumers of products and services. With ever increasing access to advance telecommunication technologies, customers are acquired more knowledge, information, and skills all the time. To deal with the informed customer, companies are exploring new approaches to marketing. There is a growing trend where business is shifting away from product-center strategies to customer-oriented strategies that are focused upon customer satisfaction and loyalty. As customer bases become more aware of the power they hold, it is increasingly imperative that companies understand the customer's perspective on products and services. The principle sources of customer value are product quality, service quality, price, brand image, and "customer–company" relationships.

There are two basic approaches for creating customer value: the effective operations and a strategic orientation. With the effective operations approach to value creation, management seeks to do the same thing of their competition only to do it better. The drawback to this approach is that no matter what the company does, their competition can easily duplicate. Thus, this approach will not yield a sustainable advantage. With the strategic orientation approach, management seeks to differentiate the company from its competition by doing different things while simultaneously transferring value to the customer. The strategic orientation approach requires a systemic process capability of dynamically developing value generating solutions for the customer. Customers can roughly be categorized into three types:

1. Customers that are interested in up-to-date modern products.
2. Customers that prefer cost-effective products and services and frequently have special requests concerning convenient purchase and high quality.
3. Customers with exacting requirements that are willing to pay a premium, or to wait, in order to get what they want.

Effective value-creating strategies must look beyond just product and services. They should also focus on the effective deployment and utilization of resources. Selling a product or service is just one way of achieving superior profits. Examples of alternative strategies are as follows: licensing, franchising, selling information, and selling excess capacity. In the traditional mind-set, management has always feared exposing any portion of the business the competition, believing that the exposure would lead to opportunism, which would result in lost market share.

Alternatively, if you research and develop a new technology, why not license it to the competition? Consider this, if you don't license it, the competition will eventually reverse engineer the technology and compete against you anyways. If you do license the technology to them, first you make money on their usage, and second, they want be interested in developing their own innovations, thus yielding you the competitive edge in new technology. The same argument can be applied to franchising why not make money on someone else's efforts? If you are collecting market data and doing analysis, then you have the first insights into trends in the marketplace; thus, you have a head start on using that information. Sell the information to your competitors. They still have to analyze it and apply the knowledge. The principle point here is that with a well-designed organization, these alternative revenue sources will provide shareholder/investor value, without seriously compromising the company's competitive position.

3.3.1 Creating Value in Manufacturing

Moving away from its roots in craft-oriented production, manufacturers have been very successful in improving productivity and quality through the adoption of strategies based upon mass production, standardization, division of labor, and control. In today's markets, manufacturing as it has been traditionally thought of is falling behind the needs of the customers constituting those markets. Manufacturing needs to evolve beyond just the production of products. There are numerous examples of manufacturing companies that do not engage in production activities, or companies that do not hold production as a core competency. Some of these companies only engage in research and development, design, and services, outsourcing their production. In USA, a software company that develops software and distributes it via a disk format is classified as a manufacturing company, while a competitor company that distributes its software via the internet is classified as a service company. Both companies are engaged in basically this same set of activities; why are they classified differently? The answer is based on how the individual firms formulated their competitive, value generating strategies. Is the value generated primarily by the services or by the products? Also, of importance is where do the revenues get generated: primarily from the services or from the products? (Fig. 3.2).

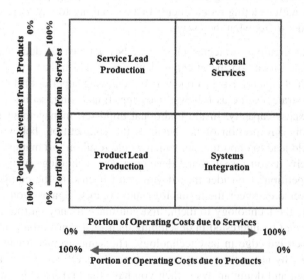

Fig. 3.2 Type of value-based manufacturing operation (Published with kind permission of © Cambridge University, The institute for manufacturing. *Source* Finbarr [1]. All rights reserved)

To be competitive in the future, manufacturing companies need to develop strategies that embrace the needs of our changing markets. The successful manufacturing model of the future is going to be based upon operations that engage in the full cycle of activities that lead to innovative new products and services. These activities are as follows: research and development, design, production, logistics, product and service deployment, and end-of-life management. These activities will all need to be provided in an economic and social context.

There are two principle challenges in creating a value proposition for the customer: first, defining and delivering a product or service that customers want and second, convincing customers to pay what the product is worth.

3.3.2 Creating Value in Services

True value in services is created by personalizing the service delivery and conveyance of an understanding of the customer's needs. Even though this makes sense and sounds easy, many companies are failing to consistently satisfy their customers. The primary reasons for this failure are rooted in two extreme paradigms of service: the skilled servitude model or the service factory model. The skilled servitude paradigm emphasizes responsiveness, customization, and empathy that are achieved through a phalanx of skilled and experienced employees, while the service factor paradigm is focused upon efficiency, consistency, and cost-effectiveness that are delivered through systems, standardization, and control.

Service is defined as "doing the work of your customer." A high level of contact, communication, and coordination with the customer is required in order to effectively deliver a service. In turn, the service provider needs to know his individual customers so as to deliver an efficient, personal, and effective service that will satisfy each customer. In accomplishing this task there are tradeoffs to be made. The service factory approach tends to treat customers as anonymous, interchangeable components within the system; while the skilled servitude approach is very expensive and hard to control the consistency of delivery from one employee to the next.

Kolesar, Van Ryzin, and Cutler suggested the service companies need to integrate information technology into their processes, thus creating systems that are capable of capturing and deploying customer knowledge. By coupling these new systems with delivery processes and organizations that are focused on providing intimate, high-value service, companies will be able to deliver customer satisfaction on an industrial scale. To accomplish this paradigm change, service providers must develop detailed knowledge about their customers, adopt the principle of Do-It-Once and Do-It-Right, enable customers to engage in value-enhancing self-service, let customers design the product, provide customers with a one-stop-shopping experience, develop competency into the delivery systems, and build long-term relationships with customers.

What creates true and equally importantly, perceived value is your ability to personalize service delivery and convey an aura of understanding and excellence to your customers.

Two paradigms of service are skilled servitude model or the service factory model. In the skilled servitude model, the focus is on responsiveness, customization, and empathy achieved through a phalanx of skilled and experienced service employees. The service factory model emphasizes efficiency, consistency, and cost-effectiveness. These objectives are delivered through systems, standardization, and control.

Reference

1. Finbarr, L. (2006). *Defining high value manufacturing, White paper*. Cambridge: Cambridge University, The Institute for Manufacturing.

Chapter 4
The Philosophy of Quality

The philosophy of quality has traditionally focused upon the development and implementation of a corporate wide culture that emphasizes a customer focus, continuous improvement, employee empowerment, and data-driven decision making. The drivers of this philosophy are rooted in the alignment of product and service systems design with customer expectations, along with focusing on quality during all phases of development, production, and delivery. The philosophy is process centric and emphasizes the reduction of variability as well as a continuous improvement in the functionality of the final product or service.

It is widely recognized that quality cannot be inspected into a product or service; it needs to be planned for, designed for, and built into products and services. If this exercise is performed incorrectly, a quality result cannot be achieved. In an investigation of the relationships between total quality management (TQM) practices, plant performance, and customer satisfaction, there was found a strong direct linkage between quality management practices and customer satisfaction. They also found that a good design helps the firm to develop and produce new products more quickly, while simultaneously minimizing engineering changes and costs.

Frederick W. Taylor was the first to advocate that the most efficient method for the accomplishment of work was to standardize work methods. He also espoused that management was responsible for the design of the processes and methods used by the firm to produce its products and services. In 1939, Walter Shewhart advanced the importance of understanding variation and the usage of the scientific method for performance improvement. Many others introduced various tools and methodologies based upon these principles; all of which led to many firms within the USA to adopt this philosophy and subsequently improve efficiencies and achieve higher productivity.

After World War II, the USA was virtually the only country with a developed and intact industrial base. Because of this situation, businesses had few worries about competitive pressures and focused their innovative efforts on optimizing internal operations. Quality was not a paramount issue. But, by the middle 1980s,

© Springer-Verlag London 2015
G.N. Kenyon and K.C. Sen, *The Perception of Quality*,
DOI 10.1007/978-1-4471-6627-6_4

the paradigm had shifted toward an organizational wide, team based, customer-focused philosophy. This change was driven by the blending of Western industrial management tools and methodologies, with Japanese social norms and innovative production techniques developed during the post-World war II rebuilding.

After the end of World War II, the Japanese industry base recovered much of their industrial base. However, Japanese industry not only recovered but also was extremely competitive in the international marketplace. Soon, the Japanese over-took the USA by offering products that were more reliable and of higher quality. The result was that US companies started losing market share. What really con-fused US managers was that not only were the Japanese products smaller (or had fewer features), but they were also higher priced.

How did this happen? After the war, the USA sent industry experts to Japan to educate their industry leaders on tools and principles of management used in Western industry. What transpired was a blending of those tools and principles with the cultural philosophies of Japanese society. The result was a new manage-ment philosophy that was based upon such principles as respect for individuals, teamwork, and extensive planning before execution. The overall result of this blending was a new management philosophy focused on customer needs and qual-ity product and services that meet those needs, and continue improvement. Even with the successes of Japanese companies in the world markets, the philosophy originally remained specific to Japan. In order to remedy this situation, quality management experts such as W. Edwards Deming, Joseph Juran, Philip Crosby, Armand Feigenbaum, Kaoru Ishikawa, and others often travelled across the USA educating American-based executives about this new management philosophy.

4.1 Foundational Theories of Total Quality Management

4.1.1 W. Edward Deming

Deming espoused that higher quality leads to higher levels of productivity, which in turn leads to increased long-term competitiveness. The main components of Deming's philosophy on TQM were his chain reaction theory, the fourteen points of TQM, and the theory of profound knowledge. Though Deming's original audi-ence was the manufacturing industries, the principles of TQM soon transcend industry boundaries: reaching into services, health care, government, and more.

4.1.1.1 Deming's Chain Reaction Theory

The application of Deming's chain reaction theory implies two basic sets of reac-tions. The first reaction is that as a company improved its transformation process-ing quality, work would increasingly be performed correctly on the first pass, resulting in lower costs because of less scrap, rework and warranty costs, fewer

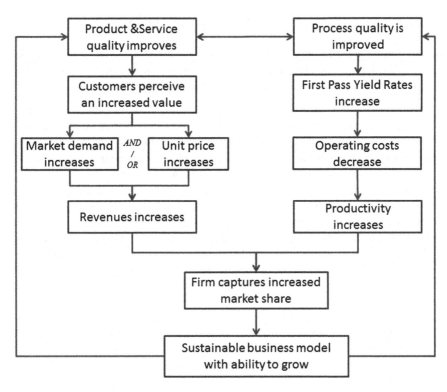

Fig. 4.1 The relationship between quality and profitability (Published with kind permission of © Cengage Learning SO. *Source* Evans and Lindsey [1]. All rights reserved)

delays and snags in material flows, and an overall better use of time and materials. This, in turn, would result in increased productivity. As the company's cost structure decreased, its profit margins would increase, leading to greater market shares. The second reaction is that as the company improved the quality of its products and services, its customers would perceive a greater value proposition that would increase market demand allowing the company to increase its pricing, or increase sales volumes, or both; thus, increasing revenues. Both these factors would lead to higher market share. The ultimate effect is the company would become more competitive within its sector and continuously extend its life and thus provide a steady stream of employment opportunities. This concept is illustrated in Fig. 4.1.

Both reaction chains are driven by top management focusing on the development and propagation of quality throughout the organization. Deming's fourteen points provide a checklist for this focus. Though decision making in both reaction chains are customer driven, the understanding of customer needs and how to achieve satisfaction is greatest in the first reaction chain because product/service design decisions must be governed by an understanding of how product features and service attributes will be perceived by the customer.

4.1.1.2 Deming's Fourteen Points of Total Quality Management

Deming articulated for management fourteen organizational objectives that he believed formed the foundation for the transformation of any organization into a total quality competitor. These fourteen pillars are as follows:

- *Create a constancy of purpose*: Management must create a constancy of purpose within the organization for continual improvement of products and service to society. Thus, all resources would be allocated to provide for long-range needs rather than only short-term profitability. This is consistent with the goal of improving competitive, staying in business, and providing jobs.
- *Adopt the new philosophy*: Management must awaken to the challenges facing us in today's world. They must learn their responsibilities to the organization and the customer and undertake leadership for change.
- *Plan for quality*: Management must eliminate the need for mass inspection as the way of life to achieve quality. Inspections are only good for sorting good output from bad and for collecting information. If the goal is to produce a quality output, it must be planned for and the appropriate philosophy built into the product creating process.
- *End the practice of awarding business based on price*: Management's focus should be on the minimization of total cost and not on the lowest cost for a single item. When long-term relationships are established between business partners several things happen. First, they start working together toward mutual goals. Second, information is shared more readily, resulting in a decrease in uncertainty. Third, work and processing activities can be aligned between different partners to better leverage organizational capabilities leading to greater value creation. The end result is that quality is continually improved while total costs are decreased.
- *Continuous improvement*: Nothing in our world stays static, everything changes. If a program or company is to stay competitive, it must improve its processes at least as fast of its competitors are changing. Management has a responsibility to evaluate how fast the marketplace is changing and to continually align (and re-align) resources to drive improvements at that rate or faster. A continuous improvement in every process involved in the production of products and services is therefore essential. The organizations must also develop their workforces to continually search for problems in order to improve all activities within the company, thereby enhancing quality and productivity and also lowering overall costs.
- *Institutional training*: People are at their most productive when they have the requisite skills and knowledge for the performance of tasks they have been assigned. As processes and methodologies are improved or changed, the workforce needs to be trained with the new skills and knowledge. Additional training on skills and knowledge on tasks peripheral to their primary tasks makes workers more flexible and consequently improves the versatility of the organization. The greater the versatility of the organization, the leaner it can run and still deliver the products and services demanded by its customers in a cost-effective manner.

- *Institutional leadership*: Leadership is an important factor in the effective directing and motivating of a workforce. Leadership is also a characteristic that can be exercised by different employees at every level. By adopting and instituting leadership aimed at helping people do a better job, a proactive organizational culture can be created where everyone in the organization will take responsibility for the processes affecting the customer. In order to build the trust needed for this outcome, management must ensure that immediate action is taken on reports of inherited defects, maintenance requirements, poor tools, fuzzy operational definitions, and all conditions detrimental to quality. At the same time, management must also be aware of competitive product in the market place and strive to ensure that their own offerings provide better value to the customer.
- *Drive out fear*: Management is responsible for the designing, monitoring, and controlling of the organizations processes, as well as the planning of future activities and organizing of resources for the carrying out of those plans. The work force is responsible for the execution of management's plans. When problems occur, as first responders, the work force will be the ones to first notice them and understand their implications. Without an organizational culture that encourages workers to instantly recognize potential problems and elevate relevant issues, management may never become aware of possible hitches, until it is too late to prevent them from affecting the customer. Thus, management needs to encourage effective two-way communication and other means of driving out fear throughout the organization so that everybody may work effectively and more productively for the company.
- *Break down barriers between functional areas*: In today's fast paced world, companies are being driven to making decisions and delivering products and service at an ever increase rate. One of the best tools for analyzing and maturing decisions that meet requirements on the first pass is cross-functional teams. In most organizations, knowledge is specific to managed function-related departments. Often bureaucratic barriers stifle the communications between these departments. This impeding of the information flow degrades decision making, which can result in unanticipated problems. Cross-functional teams (i.e., people from different areas, such as leasing, maintenance, administration, and other departments) circumvent these barriers by bringing all of the necessary people together to tackle problems that may be encountered with products or service.
- *Eliminate exhortations*: The practices of using slogans, posters, and exhortations for the work force, demanding Zero Defects and new levels of productivity, without providing the appropriate methods and resources frequently result in failed initiatives and lowered moral. These methods of exhortation generally create adversarial relationships between management and its work force by solely focusing on meeting goals that were not realistically set in the first place. Deming believed that eighty-five percent of low quality and low-productivity problems were caused by poorly designed and managed systems (i.e., area of management responsibilities) and thus are beyond the capabilities of the work force to fix.
- *Eliminate quotas*: Work standards that prescribe quotas for the work force and numerical goals for people in management rarely result in the production of quality products and services. If management has not designed its processes with

the appropriate capacity levels and quality-based capabilities, the work force will not be able to meet quotas or goals the system's parameters. Instead, management needs to replace quotas and numerical goals with aids and helpful leadership in order to achieve continual improvement of quality and productivity.

- *Remove barriers that rob people pride of workmanship*: No one likes go to bed at night knowing that the next day will be characterized by either unproductive boredom or frustration. They would much prefer to go home at the end of the day feeling that they have not just wasted 8 h of their life. Management can make this happen by removing the barriers that rob hourly workers as well as and people in management of their right to pride of workmanship.
- *Encourage education and self-improvement*: By instituting a vigorous program of education and encouraging self-improvement for everyone, management can help create, or reinforce, a continuous improvement culture as well as a culture open to innovation. Total quality-focused organizations need more than just good people; they need people that keep improving with education, because advances in competitive position are rooted in the acquisition and application of knowledge.
- *Everyone in the company must work to accomplish the transformation*: Quality is a team effort facilitated by clearly defined and permanent commitment of top management to the continuous improvement of quality and productivity along with their obligation to implement all of these principles. It is not enough that top management commit themselves for life to quality and productivity. The work force must also need to know what it is that they are committed to and that their commitment is supported and appreciated.

Only top management can create a structure that will push every day on the preceding 13 Points and take action in order to accomplish the transformation. Support this philosophy is not enough, action is necessary!

4.1.1.3 Deming's System of Profound Knowledge

In order to improve an existing system, it is necessary to back away from it and view it from an objective third-party perspective. Given that all work is accomplished through the application of a process and that organizations are essentially a system of processes, it is essential for one to develop an understanding of how a given organization truly works in order to develop theories on improving the organization. One of the key issues in theory development in the quality management field is the articulation of the distinction between quality management practices (input) and quality management performance (output). Deming's "system of profound knowledge," consists of four interrelated parts: appreciation of the system, understanding variation, the theory of knowledge, and psychology.

- *Systems*: Work within organizations is accomplished through a series of interdependent processes. Systems governance in the typical organization is aligned according to functional skill sets (i.e., departments). When interactions occur between processes, or other parts of the system, managers cannot effectively manage the system by focusing of individual parts. They must understand the processes that constitute the system as well as their cross-functional boundaries. They must then realign these processes toward organizational goals and optimize their interactions.

- *Variation*: Every process has variances in output. The degree of variability is inherent to the basic design of the process. Variability can also be incurred when factors outside the basic design interact with the operations of the process. Excessive amounts of variability will make predicting the process' output increasingly difficult. This will lead to more wastage and rework. Deming stressed that management must first understand the process and the nature of variability, and then work to reducing variations in process output through improvements in technology, process design, and training.
- *Theory of Knowledge*: Recognizing the works of Clarence Lewis, Deming stressed two underlying facts about the creation of knowledge:

  1. That knowledge was not possible without theory. "There is no knowledge without interpretation. If interpretation, which represents an activity of the mind, is always subject to the check of future experience, how is knowledge possible at all? ... An argument from past to future at best is probable only, and even this probability must rest upon principles which are themselves more than probable" [1].
  2. Experience alone does not establish theory. "Experience without theory teaches nothing. In fact, experience cannot even be recorded unless there is some theory, however crude, that leads to a hypothesis and a system by which to catalog observations" [1].

- *Psychology*: The application of psychological theories, principles, and methods helps us understand people, the interactions of people and circumstances, the interactions between leaders and employees, and a given system of management. In order for management to lead the push for quality improvement, they must be aware of the difference between the people that they are leading, and work toward optimizing everyone's abilities and performance. An insightful leader understands that people learn in difference ways and at different speeds and will manage the system accordingly. By applying psychology, leaders can nurture and preserve the innate, positive attributes of their people.

4.1.2 Joseph Juran

Is it better to continuously improve every process all of the time? What about in situations when the cost of fixing a problem is significantly greater than the cost of reworking only the specific faulty items that were produced? What if the problem is of lesser strategy value than another problem and there are only enough resources available to fix one of the problems? These situations are common in business, and if the goal of being in business is to make money so you can stay in business, then decisions must be prioritized.

Where Deming advocated major changes to an organization's culture, Juran believed in working within the organization's existing system to improve quality. Thus, his approach focused on fitting quality changes into the firm's strategic business planning process. The centerpiece of Juran's philosophy was that there was an optimal level of quality, based upon a trade-off between quality and costs. He believed that because of the nature of these costs, this optimal level would be less than 100 % product conformance to specification.

Recognizing the interdependent nature of business processes, which include market research, product development, design, manufacturing planning, purchasing, production and production control, sales, and customer feedback, Juran advocated for a company wide cultural shift. His prescription for this shift is focused along three processes, called the Quality Trilogy: quality planning, quality control, and quality improvement.

In dealing with the costs of quality, Juran's model fails to account for the effects of technological innovations and competitive pressures. Others have noted that costs tend to increase at a constant rate, or remain flat, or even slightly decline in production environments where there is a strong learning curve, automated inspections, and other new quality practices. Furthermore, Juran's assumption that appraisal and prevention costs could be combined has been questioned. It has been observed that when prevention costs increase, defects will go down causing failure costs to decrease and diminishing the need for investments in appraisal activities. Finally, it has been noted that when appraisal cost increase, external failures can decline while internal failures will typically increase.

4.1.3 Phillip Crosby

Crosby's philosophy focused upon reducing cost through quality improvement. He stressed that both high-end and low-end products could have high levels of quality. His philosophy emphasized:

- Quality means conformance to requirements.
- There is no such thing as a quality problem.
- There is no such thing as the economics of quality; doing the job right the first time is always cheaper.
- The only performance measurement is the cost of quality, which is the expense of nonconformance.
- The only performance standard is "Zero Defects."

4.1.3.1 Crosby's Cost of Quality

Critical-for-quality (CoQ) is frequently characterized as the sum of costs associated with ensuring conformance to standards and the costs associated with failing to create a quality product or service on the first pass (e.g., nonconformance). Thus, every time a defective product is produced, or a less than satisfactory service is delivered, the cost of quality increases. Examples include reworking a manufactured item, retesting an assembly, rebuilding a tool, correction of a bank statement, or the reworking of a service, such as the reprocessing of a loan operation or the replacement of a food order in a restaurant. In other words, any cost incurred as the result of failing to produce a quality item the first time, contributes to the cost of quality.

There have been several approaches to the measuring of Feigenbaum's classification of quality costs into three main categories: prevention, appraisal, and failure. The basic premise of this P-A-F model is that investments in prevention and appraisal activities will reduce failures and that continued investments in prevention activities will lead to reduction in appraisal costs. The costs categories in the P-A-F model are generally described as follow:

Prevention costs are those costs associated with all activities designed to prevent poor quality. Examples include:

- New product reviews
- Quality planning
- Supplier capability surveys and certification programs
- Process capability evaluations
- Quality improvement projects and associated team meetings
- Quality education and training.

Appraisal costs are those costs associated with the measuring, evaluating, or auditing to assure conformance to quality standards and performance requirements. Examples include:

- Incoming and source inspection/test of purchased material
- In-process and final inspection/test
- Product, process or service monitoring and control systems and audits
- Maintenance and calibration of measuring and test equipment
- Associated supplies and materials.

Internal failure costs are those costs resulting from not conforming to specifications. Here, the defect is caught prior to delivery of the product, or the furnishing of a service, to the customer. Examples include:

- Scrap and its associated opportunity costs
- Rework
- Re-inspection
- Re-testing
- Material review
- Downgrading of materials or services.

External failure costs are those costs associated with the defect being caught after delivery of the product. In these cases, an additional service has to be provided to the customer. Examples include:

- Processing customer complaints
- Customer returns

- Warranty claims
- Product recalls
- Lost sales (typically unknowable).

Accepting Crosby's definition of quality as "conformance to requirements," CoQ's can be calculated as the sum of the price of conformance and the price of nonconformance. The principle differences of this calculation form the P-A-F model is that price is defined as including all the benefits, overheads, and whatever else that is associated with the real costs of the company.

The British Standards Institute, in an effort to extend the concept of quality costing to all functions of an enterprise and to non-manufacturing organizations, published a model focused on process costs (BS 6143: Part 1, 1992). In this model, the CoQ's are collected for a specific process, as opposed to the whole company. This model pursues a continuous improvement approach to process management that is reflective of both the Kaizen approach and to Deming's (1986) plan-do-check-act (PDCA) cycle.

4.1.4 Armand Feigenbaum

Feigenbaum believed that TQM was the most effective method for integrating the various quality activities of multiple groups in an organization while enabling production and services to deliver customer satisfaction at its most economical level. Furthermore, he believed that in the typical non-TQM manufacturing environment, there was so much extra work being performed in correcting mistakes that there was essentially a "hidden" plant within the factory. Most importantly, he believed that quality was everyone's job, and without upper management actively and visibly involved, no one would do it.

Feigenbaum stressed that quality did not mean the best performing, or the best technical option. Instead, it means the best for the customer's usage. In the system that he envisioned, quality control was a four (4) step management tools involving: setting of quality standards, appraising conformance to those standards, acting when standards were not meet or were exceeded, and planning for improvements in the standards. The following tenets sum up the essences of Feigenbaum's system of TQM;

- Quality is a company wide process.
- The customer defines quality.
- Effective quality requires both individual and team effort.
- Quality is a management philosophy.
- Quality and innovations are mutually dependent.
- Quality is an ethical standard.
- Quality requires continuous improvement.
- Quality is the most cost-effective method for improving productivity.
- For quality to work properly, it must be implemented as a total system and involving both customers and suppliers.

4.1.5 Kaoru Ishikawa

Even though Deming, Joseph Juran, and Phillip Crosby are the most famous of the quality gurus, it was Kaoru Ishikawa that laid the foundations of TQM by initiating a company-wide quality control (CWQC) system. His vision for CWQC held that quality was not just about the conformance/performance of the product, but also included after sales service, quality of management, the company itself and human life. He believed that quality improvement was a continuous process that could always be taken one step further. He emphasized that quality was important throughout the product's life cycle.

Ishikawa believed that the creation of standards was important, but did not believe that they were the ultimate source of decision making; customer satisfaction was. Thus, he endorsed that standards should be constantly evaluated and changed to insure customer satisfaction.

Ishikawa is most noted for his contributions in simplifying statistical techniques for quality control through the development of simple tools for the application of statistics such as control charts, run charts, histograms, scatter diagrams, Pareto diagrams, and flowcharts. In addition to these tools, he invented quality circles and created the cause-and-effect diagram

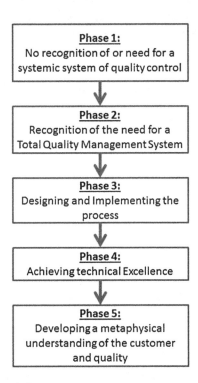

Fig. 4.2 Phases of TQM evolution

4.2 Evolving the Philosophy of Quality

The development of a TQM program is a multiphased process. As can be seen from the discussion of the foundations of TQM, there is a consistent emphasis of customer satisfaction, customer focus, and customer-driven quality. These concepts are unfortunately not well-defined in the literature in this field. Thus, most organizations are stuck in the third phase of their TQM evolution (see Fig. 4.2). In short, many of them have not developed a cogent philosophy of the quality of their final product that emanates from a deep understanding of the targeted consumer.

For an organization to reach its full potential, it is understanding of the philosophy of quality must evolve beyond a technical understanding of customer focus, continuous improvement, employee empowerment, and data-driven decision making, to a deeper understand of how customers will perceive various actions. In addition, the firm must try and ascertain what the true desires of customers and strive to deliver products and services that provide their customers with a visceral level of satisfaction.

Reference

1. Evans, J. R., Lindsey, W. M. (2008). *Managing for quality and performance excellence* (9th ed.). Mason, OH: South-Western Cengage Learning.

Chapter 5
The Perception Process

Different people tend to perceive the events around them in different ways. This tendency is a result of their personalities and their differing backgrounds and experiences. It is also a result of how we, as individuals, process various stimuli. Perception is a process where a person selects, organizes, identifies, and interprets the sensory information he receives in order to understand his environment. Perceptions act as filters, thus preventing us from being overwhelmed by all of the noise (stimuli) around us. Perceptions can be shaped by learning, memory, and expectations. Our perceptual systems also enable us to see the world around us as stable, even when the stimulus we receive is incomplete and/or changing.

5.1 The Psychology of Perception

Our perception process is influenced by both internal and external factors. The internal factors include motives, values, interests, attitudes, past experiences, and expectations, while the external factors include motion, intensity, size, novelty, and salience. It is generally accepted that there are three steps in the perception process.

Step one involves the noticing of a stimulus, also called "attended stimulus." These stimuli can be light, sound, taste, feel, or any other physical interaction with our environment that stimulates one or more of our body's sensory organs. From our world view, we select those elements of a stimulus that we deem to be the most relevant.

Step two involves the organization of the stimulus elements, called "transduction." When our sensory organs receive a stimulus from the environment, it is transformed from input energy into neural activity. This neural activity is then structured into patterns that make sense to us.

Co-Authored with: I. Newman, M.S. I/O Psych., Lamar University.

© Springer-Verlag London 2015
G.N. Kenyon and K.C. Sen, *The Perception of Quality*,
DOI 10.1007/978-1-4471-6627-6_5

In step three, we subjectively interpret these patterns based on our individual attitudes, needs, experiences, expectations, goals, values, and physical conditions, forming what is called percepts. Stimuli are always translated into a percept. Often, our percepts are formed utilizing multiple stimuli; alternatively, ambiguous stimuli may create multiple percepts.

Our interpretation of the events surrounding us are influenced by our subconscious blinders, lack of awareness, and perceived similarities, all of which can cause errors in judgment. Our recognition of an item or event, and our subsequent actions are behavioral outcomes of the perception process. Recognition is the ability to place an object into a category, while action is the motor response our body takes as a result to our perception processing. Another outcome of this perception process is our ability to acquire and retain knowledge.

5.1.1 The Features of Perception

The attributes of individual objects form the basis for our categorization of objects that we percieve in our surrounding environment. These categories are subject to a set of behavioral rules we develop over our lifetime. The features associated with the organizational process of perception are constancy, grouping, contrast, and learning.

5.1.1.1 Constancy

Perceptual constancy is defined as the ability of one's perceptual system to recognize the same object or pattern regardless of the sensory input. An example of perceptual constancy is our being able to recognize a given object under varied shades of color, light intensity, and relative distance. Some philosophers have suggested that constancy occurs because perceivers are categorizing non-constant stimuli under constant conceptual categories; while others have explained the phenomenon as our visual system misrepresenting changes in conditions as surface spectral reflectance. Constancy is basically a perceptual property of primary qualities. The shape of an object is perceived as remaining constant regardless of how it is moved around. Like shapes and sizes, colors are perceived as being constant regardless of illumination.

5.1.1.2 Grouping

The human brain has the innate ability to abstractly organize patterns and objects into groups. Several principles have been identified that describe how the mind does this grouping of objects and patterns into categories: proximity, similarity, closure, continuation, common fate, and good form. The principle of proximity holds that stimuli that are close together are perceived as the same object, while stimuli that are far apart are perceived as two separate objects. The principle of similarity holds that stimuli that resemble each other are perceived as the same object, whereas stimuli that are different are perceived as two separate objects. The principle of closure holds that the mind can perceive complete objects, even if the stimuli are incomplete. When people see an object, such as a ball, where the shape is partially obscured, they tend to see the shape as being completely enclosed by the border and ignore the gaps. The principle of continuation says that when there is an intersection, or overlapping, of multiple objects, people will perceive each as a single uninterrupted object. The principle of common fate allows people to perceive and group objects in motion. This principle allows people to make out objects in motion, even when details about the object are obscured. The principle of good form states that the mind tends to group objects based upon similarities of shape, pattern, color, etc.

5.1.1.3 Contrast Effects

Contrasting effects occurs when the perceived qualities of an object affect the perception of another object. In other words, when a characteristic of one object is extreme in some dimension, objects in close proximity can be perceived as further away from that extreme. An example of this effect can be noted when looking at people of different heights; a person of average height will appear to be short when standing next to a tall basketball player, but will not be perceived as short when standing next to a tall building. As mentioned previously, perception is a three-step process of selection, organization, and interpretation. Culture affects each of these three steps.

5.1.1.4 Perceptual Learning

Experience teaches people how to make finer distinctions in their perceptions, by providing new kinds of categorizations. By attending wine tastings, a person can learn to distinguish the various favors and characteristics of wine. Doctors learn to read greater details and information in X-rays by seeing multiple X-rays and being shown different cases. Often, reading a book a second time, the reader will find items missed on the first reading.

5.2 The Effects of Culture on Perception

Research has found that background and environment will influence our perceptions. Culture also impacts how we select and organize information and sensory data. For instance, people who live in rural areas can sense crooked or slanted lines more accurately than those who live in urban areas. When describing a scene, Japanese people will describe the scene, commenting more on the relationships among the objects in the scene, while Americans will focus more on the principle objects in the scene and less on the relationships between the objects. It has been suggested that while our cultural upbringing form the basis for how we perceptive things, extended exposure to a new culture can modify our sensation and cognitive processes. Prolonged immersion in a new culture enhances the influence of the new culturally specific ideas and practices on one's cognitive processes.

With the first step of perception (selection), it has been suggested that we do not consciously see any object unless we need it, or have an interest in it, or just want it. In those cases, we are more likely to sense it out of all the competing stimuli in our environment. Our cultural background provides us with a template of priorities with respect to various items. For instance, it is easy to overlook an animal that is lying still, but even when you are not looking at it directly, when it moves our eye draws towards it instantly. Though this reflex is conditioned into us as a survival instinct, it also happens in nonthreatening environments.

In the second step of perception (organization), we organize the selected stimuli in meaningful ways. How we categorize stimuli is influenced by our cultural background. In an experiment on how children identify objects, American and Chinese children were asked to group two of three given objects: a chicken, a cow, and grass. The American children grouped the chicken and the cow together, while the Chinese children grouped the cow and grass together. In the American culture, grouping objects by their taxonomy is more important than the relationship between the objects. East Asian cultures tend to organize information by thematic, or functional, relationships. It has also been found that older Americans use a categorization-based strategy when organizing information into memory to a greater extent than older Asians.

Interpretation is the third step in the perception process. This step provides meaning to the information we have picked up from our environment by decoding the sensed data. The cultural affect is significant in the decoding of sensed data.

People are constantly making judgments as to age, height, social status, education, and other characteristics. The values attached to the cues we use to make these judgments are often culture centric. An example of how culture influences our judgment can be seen in how people view different animals. In America, dogs are pets. We typically relate warm, loving feelings with our dogs. But, in the Arab world, dogs are seen as unclean and lowly life forms. Dogs may be used for hunting or guard, but they are not pets. In other parts of the world, dogs and cats are seen as food sources. In the Hindu world, elephants are considered holy. In India, rats are feed, but in America, they are viewed as unclean, diseased vermin.

Sensory context (i.e., high context or low context) is another way in which culture affects perception. Studies have shown that European cultures focus attention on objects independent of context, whereas Asian cultures focus on the context. Low-context cultures are defined as those in which the meaning of a message is explicitly encoded in the message, as opposed to high-context cultures where most of the meaning of a message is found in the physical environment surrounding the object of the message.

With low-context cultures, verbal messages tend to be elaborate, highly detailed, and specific. Verbal skills are highly valued in these cultures. In high-context cultures, most of a message's information is found in either the physical context around the message, or in the relationship between the people. Little information is codified, explicit, or transmitted as part of the message. In low-context cultures, people are spoken of using attributes independent of circumstances, or of personal relations, but in high-context cultures, the person in connected, fluid, and conditional.

5.3 Making Inferential Judgments

People frequently make decisions based upon limited information and knowledge. Through lessoning, observation, and experience, we collect information about some of the characteristics and benefits associated with the products and services to be purchased. Often, many important pieces of information are still missing, and we must go beyond what we know to fill in the gaps. This process of filling in the gaps is called "inference." We form our inferences by linking information using cues, heuristics, logic, and other means, and formulating conclusions. There are two basic inference processes: induction and deduction.

Inductive reasoning is the process of taking specific information and generalizing a conclusion. Induction is used in generating hypotheses and developing alternatives, learning, forming generalizations, and predictions. Inductive reasoning permits the prospect that a conclusion can be false, even if all of the premises are true. Consumers in the marketplace frequently use specific product characteristics or service attributes, brand names, or other cues in forming an opinion about the benefits of various other products or services. An example of this behavior can be seen when someone assumes that a product is of high quality because it has a high price tag.

David Hume, an eighteenth-century Scottish philosopher, argued that in our everyday lives, people will make uncertain judgments based upon relatively limited experiences rather than forming valid deductive conclusions. He also concluded that even though we know that inductive reasoning is unreliable, we would still rely on it.

Deductive reasoning is the process of developing specific conclusions from general principles and/or assumptions. Deduction is associated with evaluation, logical reasoning, and diagnosis. While inductive reasoning allows for a false conclusion based upon valid premises, deductive reasoning demands true conclusions when the premises are valid. It is believed that deductive reasoning is a skill that we develop naturally as opposed to requiring formal teaching or training.

5.3.1 Inferential Biases

Both inductive reasoning and deductive arguments are subject to bias. Bias will distort our conclusions, thereby preventing us from forming logical conclusions. The three promenade types of bias are availability heuristic, confirmation bias, and the predictable-world bias.

The bias of availability heuristic causes us to form conclusions using information that is readily available. It is basic human nature to keep things as simple and straight forward as possible. Most people do not like the unknown, and complexity is perceived as leading to unknown results. Thus, people will have a tendency to rely on information that is readily available in their respective environments. For example, if someone asked you whether your college had more students from Austin or Houston, you would answer the question based on the relative availability of examples of Austin students and Houston students based on the availability heuristic. If you recall more students that come from Houston, you will be more likely to conclude that more students at your college are from Houston than from Austin.

Conformation bias occurs because of our tendency to confirm a hypothesis rather than deny it. It has been shown in multiple studies that people are prone to solutions that are consistent with current thought and beliefs rather than attempting to refute a hypothesis to the contrary. For example, when watching the news, people prefer sources that affirm their existing political attitudes. They also tend to interpret ambiguous evidence as supporting their existing position.

The predictable-world bias is based on our tendencies to see order in situations where order has not yet been proven to exist. The premise of this type of bias arises from wishful thinking and/or an aversion to conflict. One example can be seen in gambling, where an individual believes that he sees a pattern in a random process and attempts to predict the next result. It is difficult for some people to accept that their perception of order is different from what they are actually experiencing.

5.4 The Expectation Theory

Expectancy theory holds that a person will behave or act in a certain way because they are motivated to select a specific behavior due to what they expect the result of that behavior to be. An individual's motivations are derived from a process that governs choices among competing alternative activities. These choices are based the individual's estimates of how well the expected results of a given behavior will match up with the desired results. According to expectancy theory, employees would put forth more effort if they believed that effort would translate into high levels of performance, and higher performance would lead to valued outcomes. Therefore, if high levels of job performance ultimately lead to desirable outcomes, employees should be most satisfied with their jobs when they perform well and are rewarded for it.

Expectations will create what are called "perceptual sets." There perceptual sets can be either long term, such as a special sensitivity to hearing one's own name, even in a noisy room, or they can be short term, such as noticing food smells when one is hungry. A person's motivations will also create perceptual sets. Often, people will interpret ambiguous information or figures to see or hear what they want too.

Perceptual sets typically reflect the personality traits of the individual. Often, what people notice in others are those characteristics that they find as either own strengths or weaknesses within themselves. Theory holds that perceptions are not simply developed within a bottom-up process; instead, our brains use what is termed "predictive coding." This process starts with a broad state of constraints, or expectations, and as those expectations are validated by experience, more detail is added to our perceptual coding, shaping our beliefs, thus allowing us to form predictions.

5.4.1 The Trichotomy of Perception

In the marketplace, consumers buy products like cars, or services like insurance. In the process of making these purchases, they will make choices by comparing price and quality among the available alternatives. The choice between variants is influenced by the availability of information. The degree of uncertainty concerning the different quality characteristics defines a product's quality or the attributes which define a services quality. In making these choices, the consumer can reduce his risk of making a poor choice by consulting experts, or by choosing to buy from well-respected suppliers.

The expectations paradigm advocates that individuals will measure product or service performance against predictions they have made in advance of their consumption of those products or services. The higher the expectation for a given product or service choice, relative to its actual performance, the greater the resulting degree of dissatisfaction. Theory holds that in questing for information about the quality of goods or services, consumers can search for relevant information, or they can use experience to determine the quality of brands by purchasing different brands and using them. In extending this theory on information processing, researchers have noted that products and services have a combination of perceptual properties: credence, search, and experience.

5.4.1.1 Credence Properties

Credence properties are those product characteristics, or service attributes, that cannot be discriminated even after the product, or service, has been purchased and consumed. Customers frequently must rely on either information from a credible source, or make inferences from past experiences with similar items. One reason for the difficulty customers have in evaluating credence properties with previously untried products or services is that they do not have enough knowledge or expertise to make an accurate evaluation. Examples of these types of products or services are insurance, surgical procedures, automobile maintenance work, etc.

When we factor time into our conscious assessment of credence, we must distinguish between the two characteristics of credence: manifest versus latent. Manifest credence characteristics influence the buying behaviors of large subset of consumers, while latent credence characteristic do not immediately influence buying behavior but can re-emerge at a later time as an important element in a buying decision. It must be noted that a characteristic becomes manifest only when there are a variety of products, or services, possessing the same characteristics in the marketplace. Should those characteristics be elevated based upon a set of general market standards, consumers will tend to forget about it; just as they will should those characteristics cease to exist in the marketplace.

There are also several categories of credence found within both the manifest and latent characteristics such as hidden, stochastic, and bundled. Hidden credence characteristics are not detectable by inspection because they are associated with the production process and have little or no observable influence on the objective characteristics or the goods or service purchased. The hidden credence characteristic often concerns the "ethical" aspects of a production process. In the processing of food, only stringent control systems can assure specific outcomes; thus, if the hidden credence is latent, consumers will take the desired outcome for granted. An example of hidden credence characteristics can routinely be seen in the United States where consumers frequently believe that the slaughter of cattle, pigs, and other animals is relatively quick and painless. Another general assumption is that food processing systems are safe and free of contaminants and harmful chemicals.

Standardized credence characteristics are the minimum standards to which a given good or service should conform to, and to which the consumer cannot control. Frequently, we see manufacturing labels stating that the product is "Lite" or "Low Fat" or "all-natural", but for the consumer, to test these products is practically impossible. Furthermore, it is highly improbable that a consumer will know all of the standards associated with the claim.

Stochastic credence characteristics originate from multiple experiences with a type of good or service, and become credence characteristic because the consumer draws conclusion from the distribution of outcomes from those experiences. With many of the choices consumers face today, there are multiple substitute, or competing, products to choose from. When a brand, or trademark, is associated with one or more of the choices, the consumer may feel confident that the variability between products offer by the brand will be less than the variability between the choices offered from different sellers. Thus, good experiences with other products from a given brand will transfer as a positive credence over to the product in question based upon this assumption of stochastic behavior.

Finally, bundled credence characteristics emerge when a service provider offers expert advice relevant to the service rendered.

5.4.1.2 Search Properties

Search properties are those characteristics of a product, or the attributes of a service, that are easily discovered, evaluated, and compared by a consumer prior to the purchase of said product or service. Examples of this type of objective assessment can be found in the evaluation of a car's performance (e.g., speed, capacity, and serviceability) or the difference in benefits between services rendered by various doctors. Search properties include such attributes as color, style, price, fit, and smell.

Consumers are continually faced with choices in the marketplace. With today's global economies, there are even more products, brands, and services to choice from. Frequently, these choices are being made with less than perfect information. Often, information about quality differences is much harder to obtain than information about price differences. Furthermore, there exists the probability that the information is inaccurate.

To remedy the poor information dilemma, consumers obtained information about price and quality through search. Prior to the Internet, consumer seeking information had to resort to purchasing it from third parties. Today, the Internet has significantly reduced the search cost for information. There are two restrictions to search: (1) The consumer will need to inspect the options, and (2) these inspection must occur prior to experiencing the option.

Searchable information about a given product or service is generally generated by the seller or by third parties seeking to sell the knowledge they have gained from experience as information. Misleading information may get put out by these agencies, but when consumer experience does not match the searched information, the responsible party's reputation will be diminished and future sales will decrease; this effect is called "consumer power." Honest information will generate more sells than misleading information.

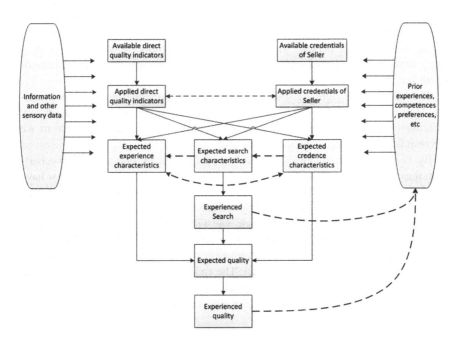

Fig. 5.1 The quality perception process model

5.4.1.3 Experience Properties

Experience properties are those product characteristics, or service attributes, that can only be evaluated during and after the purchase of the product, or the actual consumption of the service. Even though searchable information describing the product or service may be available, customers will ultimately realize that they must actually experience the product or service in order to evaluate them. Examples of this property include such qualities as taste, wearability, and purchase satisfaction.

After purchasing and experiencing a product or service, the consumer will be fully informed about price and quality for that option and thus can make a posterior estimate of the utility of the purchase. In order to know whether he is better off or not as a result of his choice, the consumer must either obtain more information on the alternative choices or purchase and experience them.

When we are exposed to information, such as when we are at an electronics store or shopping online, we receive data, one of which is sensory. Available direct quality indicators, such as product reviews and price amount, lead us to use this information and apply it as direct quality indicators. In other words, with the information we have taken in, we will now apply and develop an expected experience and its characteristics, or what we expect out of this experience of shopping at the store or online. Collectively, this information will then shape and mold the quality that we expect out of the product that we are shopping for (Fig. 5.1).

Just like sensory data that we take in as we shop, prior experiences, or knowledge that we already possess about the product that we are looking to purchase, will also affect our expectations. Likewise, the impression that we have of the individual, i.e., the sales person, who is presenting the product to us will lead us to apply those perceived credentials and then develop an expectation that we have of his or her credentials.

The left- and the right-hand sides of the flowchart operate on the same principle. However, the left-hand side focuses on quality expectations based on sensory information input, while the right-hand side focuses on credence expectations we develop from prior experiences.

By combining the quality expectations (the left-hand side of the flowchart) with the credence expectations (the right-hand side of the flowchart), we now have developed our own expected search characteristics! These characteristics will lead us to the idea of what we are expecting to find when we begin our search for the product (middle of chart). Furthermore, we will now develop an expectation of quality, which will influence the actual experienced quality. Finally, we have experienced a quality based on all the information that we have collected throughout the process of shopping for our product. The end result of this model (experienced quality) will now become a new old experience that will feed back into our storage bank that we will use for future purchases and experiences.

Chapter 6
The Aspects of Quality

In the philosophy literature, it has been put forth that quality is a naturally occurring stimulus in our environment. Every time we see something new, such as the new Ford GT, 550 horsepower, mid-engine two-seater sports cars, or experience a joyful moment, such as a well-acted performance at a play, we are connecting with this stimulus at an emotional level. Our first response is usually, "Wow!" It is only later that we are able to describe the experience in a rational manner. Our initial impression of the quality of the item or the actual moment of interaction with the product or service is termed "perceived quality" (also called dynamic quality by some philosophers). This is essentially our instant judgment about a product or service's overall excellence. Our later ability to describe the quality of the product or service in measurable terms is called "objective quality" (also called static quality by some philosophers). This aspect of quality refers to the technical excellence of the product or service.

Whenever someone organizes a company or designs a product or delivers a service, they are creating an artifact that will interact with others in hopes of stimulating their satisfaction. When this interaction stimulates a strong sense of well-being, it is described as quality.

Quality is a multidimensional concept; as such, it cannot be easily defined or measured. Because of this multidimensionality, there are many definitions for quality and several theories on the management and control of quality. Our pursuit of quality exists on two levels. As individuals, we pursue quality for the self-enlightenment and enjoyment value that it brings us. In business, on the other hand, we pursue quality because of its direct impact on customer satisfaction and the resulting competitive advantage that improving customer satisfaction can generate. However, to consistently stimulate a strong sense of well-being with all of one's customers, a greater understanding of the various aspects and dimensions by which people perceive and measure quality is required (Fig. 6.1).

© Springer-Verlag London 2015

G.N. Kenyon and K.C. Sen, *The Perception of Quality*,
DOI 10.1007/978-1-4471-6627-6_6

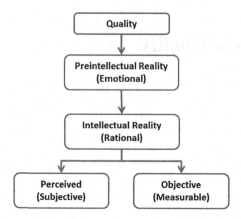

6.1 The Five Aspects of Quality

It has been suggested that the importance of a company's ability to produce and deliver quality is not so much with the improved management of the firm, or of its processes, but in the measuring and realization of quality focused improvements in products and services. When done correctly, these improvements should improve the firm's overall competitive performance. It has also been argued that there are several different aspects to quality. These cover the organization's culture and its management of quality, the quality of products and services, and finally the management and improvement of manufacturing processes.

Our knowledge of quality has developed along three distinct themes: theory and practice, characterization, and measurement and analysis. A large component of our knowledge of quality management has been focused in the area of measurement and analysis (or static quality). What is still lacking is a clear linkage between these performances metrics and the various dimensions of quality that customers use in evaluating the products and services offered to them. Furthermore, many of the practices used to control quality do not even measure all of the dimensions that define quality. In fact, most companies and control systems still treat quality as a unidimensional construct, where a single measurement defines all. In reality, quality is a multidimensional phenomenon requiring several metrics to properly measuring its affect. To complicate matters even further, these measurements cannot be used as a guaranty of success because the individual customer is the final arbitrator of quality.

The dimensions that define quality are captured within the various attributes or characteristics which are designed into the product or service. When viewed together, perceptions of performance of these attributes form the basis upon which an overall opinion of the product is formed and expectations for similar items are shaped. It should also be noted that there is a degree of interdependence between the various dimensions. Thus, any opinions or perceptions that are formed without

considering all the applicable dimensions of an item will be deficient. Thus, leading to false expectations and ultimately customer dissatisfaction.

Garvin asserted that the only way to achieve a competitive advantage through
quality is to match the importance that markets assign to the individual quality
dimensions for a product or service to the organization's ability to performance
along those dimensions. In other words, companies need to not only understand what customers want and expect, they also need to fully understand what
their own capabilities are for delivering product and services that perform along
the dimensions that define the customer's expectations and desires. Furthermore,
product designers need to have an appreciation for how the various features and
attributes of a product or service will be perceived by the target group customers.

From the customer's point of view, the value of a given product or service is
related to their expectations for the product's performance or a service's benefits,
as compared to their hypothetical ideal. Confounding the issue, each customer's
background and experiences will shape, or bias, their expectations. Thus, the
design process is complicated by the need to understand how various social, cultural, and experiential factors influence the perception of a given quality dimension. The goal is to derive a set of static specifications that conform to the broadest
cross-section of customers' expectations.

A customer's perception of a product or service will have a combination of search,
experience, and credence properties. Search properties are defined as those attributes
that can be evaluated by a consumer prior to purchase. In contrast, experience properties can only be evaluated after the purchase and actual consumption or use of the
product or service. While, credence properties are those characteristics that cannot be
easily discerned even after the product or service has been purchased. By understanding how to manage these properties, firm's gain the opportunity to shape customer
expectations and thus bring them more in line with product and service offerings.

The consumer's judgment about the quality of a product or service is influenced
by the incidence of these three properties during the product or service purchase process and usage. Many of the purchases we make can be characterized by both tangible and intangible attributes. For example, while a fast food chain serves hamburger,
fries, and drinks (a tangible product), our evaluation of the chain is also affected by
intangible factors such as the outlet's cleanliness, the salesperson's friendliness, and
the speed of service. Although the quality of many tangible products can be judged
prior to purchase (a search property), the quality of most intangibles (e.g., the service) can only be judged after purchase (an experience or credence property). Given
these differences, it is important that the various dimensions of quality from the perspective of product/service/consumer interactions be analyzed in the proper context
for their impact on the customer's final perceptions of quality. It must also be noted
that the attributes associated with the firm's organizational structure and transformational processes must be aligned with the experience the firm wishes the customer to
have in order to design and produce quality in its products and services.

An important concept regarding quality is that it is an emotional response to
the fundamental elements of a product or service. The characteristics that are built
into a product or attributes that are the components of a service generate benefits

along several dimensional lines that are perceived by the customer. The customer's perceptions are shaped by various elements of his/her environment and past experiences. Thus, quality exists at the aggregate intersection of these dimensions and perceptions. If the perceived quality of an item is unique in definition to the individual experiencing it, then the quality is dynamic in nature as shown in Fig. 6.2.

In order to maximize the customer's perception of that quality, designers must plan it, organize for its delivery, implement and execute the plan with a focus on maximizing the quality experience, and delivering the product or service in a state that assures the desired customer reaction. To accomplish this, a set of static specifications must be created that describe the product characteristics or service attributes which will elicit the same perception from the widest cross-section of customer's possible, as shown in Fig. 6.3. Furthermore, it must be recognized that there are several distinct aspects to this creation and delivery process that must

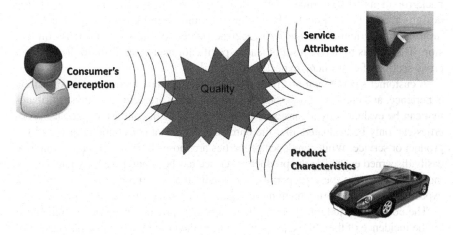

Fig. 6.2 Dynamic quality

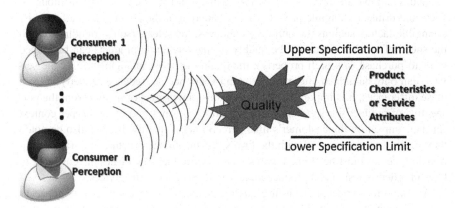

Fig. 6.3 Static quality

be managed in order to provide the customer with the intended quality experience that can yield a competitive advantage.

The first step in this process is for the firm's management is to create and maintain an organizational culture that is conducive to this level of design effort. Top management must provide the leadership necessary to focus the organization's resources on creating and maintaining the capabilities necessary for assuring customers a high quality experience. This requires that both management and employees understand how the company and its customer will benefit from improved quality. It is essential that management have a vision of how the company will mature and grow into being an organization that values quality. This entails that the company values and demands a higher standard of performance from its employees but also values its customers enough to learn what they desire. This attitude should lead them deliver only the products and services to customers that generate the highest quality.

Companies can strive to become this type of organization in four incremental steps or stages. During the first stage, quality is not a priority of the organization. Because of this lack of emphasis by management, there are no systemic standards and processes for achieving or measuring quality. With regard to their customers, the organization is not focused on their satisfaction.

In the second stage of quality maturity, the organization is focused upon avoiding mistakes and reducing waste. However, the main focus is on output instead of process and uniformity. At this stage of quality development, the organization sees quality as a problem to be solved; thus, there is a strong tendency toward inspection. Management's customer orientation is centered on the avoidance of dissatisfaction, which results in a significant level of complaint resolution and order accuracy.

In the third stage of quality maturity, the organization's culture is to aggressively attack quality problems by proactively avoiding them, instead of reactively correcting them. Management also takes responsibility for ensuring quality throughout the organization. This type of organizational culture is focused on the quality standard of "zero defects." With this type of focus, when a problem occurs, effect is made to identify and remove the root causes of the problem. This type of culture works hard to obtain customer preferences prior to designing a product or service and monitors their customer satisfaction continuously.

The last stage in the quality maturity progression is to approach quality from a creative perspective. The entire organization's strategic focus must be centered on quality and customer satisfaction. Every effort is made to design products and services that not only are defect free, but elicit an emotional response; thus, delivering the customer unexpected benefits. To accomplish this level of service and satisfaction, the organization focuses on generating lifelong loyalty among its customers through the creation of new and higher levels of performance.

If a company's objective is to create and maintain a competitive advantage, its culture must be focused on satisfying all of its customers, both internally and externally. To accomplish this goal, the consumer's needs and expectations must be understood. This includes an understanding of the environment in which these needs and wants are to be fulfilled in.

Armed with this knowledge, products and services can be designed with the features and functionality necessary for meeting and exceeding the consumer's expectations. To ensure that the functionality of products perform properly or that the benefits of offered services are realized, the appropriate transformational processes must be developed to produce the relevant features, within the required specifications on the first pass through the system (i.e., do it right the first time).

Additionally, products and services must be delivered to the customer in the required quantities, at the specified locations, at the designated time, and in the expected condition to insure no loss of satisfaction by the customer. Finally, the appropriate aftermarket support and follow-up processes must be established and maintained. Very few of these activities are visible to the consumer. Realistically, external customers do not care about the internal mechanisms but would like to be assured that the delivered products or services work as expected (Fig. 6.4).

With respect to the two principle customer groups, the company's quality improvement efforts can be divided into two principal areas of influence: quality management and quality results. The quality management areas consist of the supply chain quality, the transformational process quality, and the organizational quality. Quality result areas consist of product quality and service quality. In each of these areas, the aspects are comprised of multiple dimensions. The sum total of and interactions between the different dimensions define the essential quality

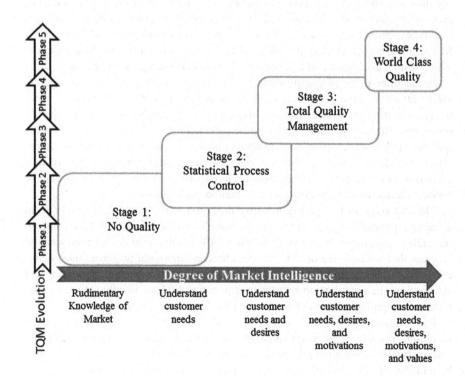

Fig. 6.4 The stages of quality maturity

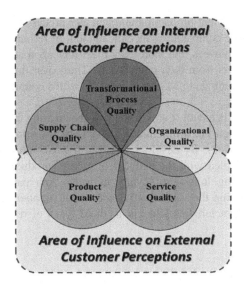

Fig. 6.5   The aspects of quality

inherent in the product and/or service experience. In fact, the consumer's perspective on a product or service's quality or the firm's capability to produce quality in general is typically based upon a synthesis of several characteristics, rather than on a single attribute.

Figure 6.5 illustrates the interdependent nature of these quality aspects and their related influence on the customer. Given the dynamics of the typical business environment companies operate in, the supplier must always be monitoring and improving these aspects in order to successfully compete in the market place.

6.2 Quality Management and the Customer

The aspects of quality that relate to its management are organizational quality, transformational process quality, and supply chain quality. In today market environment, customers are taking a greater interest in the infrastructure of companies. Because of the increased competition in virtually every market, the decision on whether or not to do business with a company is increasingly becoming about more than just about the quality of their products or their service offerings.

Two emerging trends are developing within all marketplaces as a result of globalization: (1) Customers are becoming increasingly more interested in the benefits associated with the overall relationship between themselves and their suppliers, and (2) companies are also finding that it is getting more and more difficult to differentiate themselves from their competition. In addition, there are two primary factors relating to the increased difficulty of companies differentiating themselves from their competition. The first factor is that for any given design, there is usually only one optimal production and delivery process. The second factor is that as more and more companies reduce the amount of resources at their disposal to the training of employees, the more they must rely upon acquiring skilled and knowledgeable workers from their competitors. Thus, over time, competing companies will eventual migrate to a common process with a common knowledge base for decision making.

As a result of this behavior, the best method for creating a competitive advantage will come through the differentiating effects of investing in customer relationships and interactions. By developing and/or improving the relationship with one's customers, coupled with the delivery of quality products and services, the probability of creating long-term loyalty between both parties increases significantly. In developing the value that is inherent in supplier–customer relationships, four basic characteristics inherent to these types of relationships must be understood:

1. value is subjective,
2. value is seen as a trade-off between benefits and sacrifices,
3. benefits and sacrifices are multifaceted constructs, and
4. the perception of value is relative to competitive offerings and prior experiences.

The latest trend in the creation of customer value through their joint relationship is in the constructs of economics, strategic, and behavioral value. When companies improve their quality management through the refinement of the dimensions that define the aspects of quality, customers should perceive an increase in value. When the company uses its resources efficiently in providing new and innovative solutions to customer needs and requirements relative to their competition, customers should perceive value. Customers will also perceive value from the network of processes beyond the dyadic supplier–customer relationship. There are five dimensions that define relationship value: products, services, relationship benefits, price, and the cost associated with the maintenance of the relationship. As discussed in previous chapters, value is often defined by the perceived benefits versus the sacrifices or costs. Thus, the dimensions of product, service, and relationship benefits are associated with the benefits to the customer, while the dimensions of price and relationship cost are associated with sacrifices made by the customer.

In another study, four dimensions were identified as being directly related to supplier–customer relationships (e.g., cost, quality, volume, and safeguards) and four dimensions that were related to the suppliers potential for creating value in the relationship (e.g., market, scout, social, and innovation). In measuring and monitoring a supplier's performance, both objective and subjective measures can be used. The objective, or quantitative, measurements include such factors as delivery performance, quality performance, and supplier cost reduction. The subjective measurements include supplier's problem resolution ability, technical ability, progress reporting, corrective action response ability, cost reduction ideas, product (new and existing) support, and organizational/cultural compatibility.

6.3 Quality Results and the Customer

The aspects relative to quality results are product quality and service quality. Though customers may be aware of the quality aspects of a firm's organization, their transformational process, and/or their supply chain systems, they are not going to be satisfied and may even defect to your competition for their next purchase, if the actual performance of the product or benefits of the service do not live up to their expectations and needs. The dimensions by which customers perceive and measure the quality of the products and services they purchase are discussed in the following two sections.

6.4 A Framework for Quality

The marketplace and the environment in general is the place where customers will consume services and utilize products. The ability of the product/service to robustly render the expected benefits sought by the costumer contributes significantly toward the customer's perceptions of quality. Information received by the customer about the performance of substitute products/service performance also contributes to the formation of perceptions. In the quality management literature, there are strong endorsements for companies to seek out these expectations and perceptions as the starting point for quality-based product/service design. As such, in this global economy, the international environmental conditions not only drive decisions and actions within the firm's organizational quality management processes and structure, but also influence the customer's satisfaction derived from recently purchased products and services. Conversely, the firm's ability to continuously produce superior products and services will influence the customer's perceptions of the company and satisfaction with its products and services.

Before a new product or service is developed, the firm must make a decision that there exists a profitable market for it. This leads the company to take the initial actions to start the product/service design process. This process is executed within the domain of the firm's organizational quality management processes. Thus, the firm's organizational quality practices will significantly influence the process of creating and maintaining product/service quality. Similarly, these quality practices will impact the design and maintenance of the firm's production processes and service delivery processes. Furthermore, the capabilities of the production processes and service delivery processes must be balanced between the interdependences of these processes, the requirements of the product/service designs, and the scope of the firm's competitive strategies.

Given that typically, 60–80 % of the inputs to the transformation process come from suppliers, it stands to reason that the quality of those inputs will significantly affect the quality of the firm's outputs. In fact, there is an existing body of literature on the select, development, and management of suppliers for this very reason.

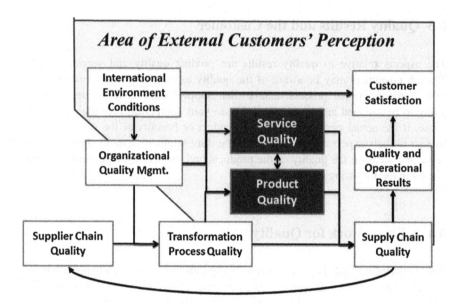

Fig. 6.6 Quality framework model

Well-designed processes typically produce high yields of products and services that conform to both specifications and the customer's requirements. This results in both high levels of quality and operational performance and customer satisfaction. Finally, the firm's supply chain functions can influence customer satisfaction through its ability to deliver quality inputs to the firm's transformational processes and to deliver complete orders in a timely fashion to the desired customer locations.

When one starts to design either a product or a service, it is important that all of these aspects of quality and their interdependences be considered. The flow path for understanding the relationship of the various aspect of quality on the product/service design process is shown in Fig. 6.6.

Reference

1. Bartneck, C. (2009). Using the metaphysics of quality to define design science. In *Proceedings of the 4th International Conference on Design Science Research in Information Systems and Technology*.

Chapter 7
Organizational Dimensions of Quality

It is widely agreed that in order for customers to recognize the inherent quality of a product or service, it must be planned for and built into the product or service. This implies that each member of the development team, and its supporting organization, must understand quality and the importance of their decisions, not only on the costs of those decisions, but also how the results of those decisions will affect the customer's perceptions of quality.

While the transformational process quality provides an important underpinning for the dimensions by which consumers judge a product's or service's overall quality, ultimately, the continued success in meeting customer's quality expectations is driven by the organization's culture. Companies not only need to understand what their customers want, but also how they are going to use the products and services offered. They must also anticipate how their competitors are going to respond to these same market demands, and then, craft solutions that customers will find greater value from. Hence, as a prerequisite to the creation of high-quality goods or services, there must exist an organizational emphasis on the customer and on quality.

The values espoused by an organization not only shape its character, by enabling each member of the organization to define their reality, but they also drive the way in which tasks are accomplished. Fundamentally, the values and beliefs that are the foundation of an organization's culture resonate through its philosophies and polices. This in turn influences the development and execution of quality management practices.

Total quality management (TQM) is an integrated, interfunctional management philosophy for achieving and sustaining a competitive advantage. It provides the core quality principles and practices that lead directly to improved performance. Moreover, it creates an environment that supports the effective implementation of these core practices. Within the TQM philosophy, there are six cornerstone values or practices that drive high quality results: customer orientation, organization-wide

© Springer-Verlag London 2015
G.N. Kenyon and K.C. Sen, *The Perception of Quality*,
DOI 10.1007/978-1-4471-6627-6_7

participation, continuous improvement, management by facts, process orientation, and leadership commitment. These values drive seven critical practices that are essential to an effective quality program: leadership, strategic planning, customer focus, information and analysis, human resource management, process management, and supplier management. In addition to these seven critical practices, innovation has also been found to significantly improve performance. Other factors found to have a significant influence on the customer's perception of quality within an organization are as follows: past quality performance, the degree of industry-wide competition, and the extent of government regulation.

7.1 The Leadership Dimension

In his 14 points for management, Deming charged top management with

- Creating a constancy of purpose toward improvement;
- Adopting the new quality centric philosophy;
- Institute leadership;
- Drive out fear;
- Eliminate slogans;
- Eliminate management by objectives (i.e., quotas);
- Eliminate the barriers that prevent employees from taking pride in their workmanship;
- Break down the artificial barriers between departments;
- Establishment a culture that values education and self-improvement; and
- Make everyone in the organization responsible for the success of the transformation.

Due to the breadth of these practices mentioned above, it is essential that top management takes ownership of the TQM initiative and provides the leadership necessary to drive its acceptance throughout the company. The role of a leader is to establish a unity of purpose and direction across the organization. This will create a culture and atmosphere where people can become fully involved in achieving the organization's objectives. There are several basic reasons for why top management assumes this role:

1. they control all of the firm's resources;
2. they set the goals and objectives for the organization, and
3. they establish the recognition and reward policies for the organization.

In addition to these key control factors, the firm's executive management must also ensure that all policies and procedures are aligned with the current goals and objectives of the organization. The degree to which this alignment occurs reinforces the stated objectives and strategies of management. It also provides structure to the decision-making process in the allocation of resources. Furthermore, it reduces the number of continuous quality improvement activities that are necessary and assists top-level management in assessing implementation efforts and performance against objectives.

Traditional thinking about leadership suggests that leaders are virtually interchangeable. One leader is not more valuable to the organization than another. The assumption is that if a pattern of behaviors that cause the organization to prosper exists, this bundle can be duplicated to ensure success in all situations. However, this approach to leadership merely emphasizes the tactical aspects of leadership, i.e., objectives, behavior, outcomes, and the measurability of effectiveness. As such, it totally ignores the importance of strategic leadership.

What is needed for organizations to create and maintain a sustainable level of competitive performance over time is for its leaders to establish a balance between tactical leadership and strategic leadership. Whereas tactical leadership focuses on values such as efficiency, specificity, rationality, measurability, and objectivity, strategic leadership values focus on holistic purpose, goodness, importance, and long-term quality.

The tactical prerequisites for leadership involve the development and maintenance of basic leadership competencies, such as, mastering the various leadership theories, conflict management tactics, team management principles, shared decision making, and group process techniques. In order to move beyond simple good leadership toward superior leadership involves the development of strategic requirements.

The first requirement is perspective: The ability to differentiate between tactical and strategic issues and understanding how they relate to each other. With perspective, leaders will bring a broader, long-range view to his/her responsibilities. It also enables one to sort out the trivial from the important.

The second requirement is principles, which provide the integrity and meaning to one's leadership. A leader's principles communicate to his followers what he values as important. As one of his/her first tasks, a strategic leader must craft a platform from which he can articulate these principles into an operational framework within which the organization can function. This platform communicates the standards and criteria by which decisions can be made.

Good leaders must develop strong political skills and awareness. Defining leadership as the ability to influence other people's behavior, and "power" as the ability to act to achieve a desired goal, it stands to reason that power is an essential ingredient in the exercise of leadership. Thus, strong interpersonal and political savvy are important in an organizational environment which invariably has multiple agendas and personalities.

In an organizational setting, it is important that leaders provide people with purpose for the goals and objectives they need to meet. Purpose provides people with a foundation for interpreting the significances of their contributions. Thus, ordinary events and actions can become meaningful and engender motivational benefits.

Few if any long-term, sustainable actions happen by accident, they need to be planned. Planning provides the structure and design for how various activities create success by laying out what actions must occur and their corresponding sequencing and performance parameters.

Another key characteristic of good leadership is persistence. This characteristic involves the consistent communication and/or signaling of those principles, issues, goals, and outcomes, which are to be valued. The importance of this characteristic is that it tells others in the organization what the priorities are. The old adage that "actions are more important that words" hold true. If the leader's actions do not consistently reinforce his stated goals, objectives, and priorities, others in the organization will start ignoring what is said and doing what is rewarded.

Few sustainable actions can be accomplished by any organization without people. A good leader continuously works to grow and develop the human resources of his/her company and to match their skill sets to the goals and objectives of the organization as well as matching tasks to the individual desires of the people. Research has found significant links between the satisfaction and development of an organization's workforce and its performance. When all of the above skills, requirements, and antecedents come together, it can be seen that leadership is less about behavioral style and more about culture.

7.2 The Strategic Planning Dimension

Far too often, management is in such a hurry to find a solution to a problem that they frequently bypass analyzing the facts before making a decision. Often, finding that their solution failed to solve the problem and that they must not only undo the negative effects of their decision, but also they must work toward solving the problem again. Compaq Computers (now part of the Hewlett Packard Company) spent untold millions of dollars trying to find an answer to Dell Computer's online marketing. Every time, management tried to tackle this problem, and they pushed people for a quick solution sighting that they were losing marketing share and needed an answer right now, only to fail again, and again, and again. The lesson that the Japanese taught us when they started expanding their markets internationally, obviously did not stick. That lesson was to take the time necessary to study the problem and plan the solution thoroughly and then push for a rapid implement.

The problem with making plans is that the further out into the future you go, the less confidence you have that events will work out the way you expect them too. On the flip side though, organizations that look to the future and plan for those expected events, as well as the probability of unexpected events, have a greater change of surveying than those organizations that do not have a road map to the future.

Strategic planning has its roots in both political and military history. The word "strategy" comes from the Greek work "strategia" meaning general or commander. Strategy is defined as a high-level plan for achieving one or more goals under conditions of uncertainty. It is a process used to prioritize and focus the efforts of an organization in the future by anticipating changes in the business environment and positioning resources to respond to those changes.

Establishing and maintaining a TQM program requires a fundamental change (i.e., a paradigm shift) in how an organization is managed and how it thinks. This is not a process that can happen haphazardly and be successful; it must be planned. In creating a strategic planning process for TQM, the organization must analyze and evaluate the needs of their customers to determine the best approach for creating value for them. The more difficult the acquisition of the resources necessary to create that value proposition, the further out into the future that planning needs to be made in order to be prepared for it.

Joseph Juran stated that strategic quality planning is a systematic methodology for defining the long-term goals of the business. He believed that this level of planning needed to include goals for the improvement of quality and the means to achieve those goals. Effective strategy plans are best developed through a process that is inquisitive, expansive, prescient, inventive, inclusive, and demanding. Key elements in the successful implementation of strategic plans include the following: the active participation of top management, the communication of planning details to all stakeholders, the alignment of organizational resources with the strategy, execution of the strategy, and the measuring of performance against the plan. Overall, the organizations top management must have a clear vision of what their desired end state looks like. They must provide direction throughout the process and empowering stakeholders to make the changes.

Strategic planning is an ongoing effect. Even if the organization has a solid plan in place, changes in the competitive environment, or even internal changes, can have a significant effect on the viability of a long-term plan. Just in the last 30 years, there have been tremendous changes in virtually every aspect of our lives. There have been technology changes in telecommunications and computers that have significantly affected everything from the geo/political to our social environments. There have been cultural changes as a result of aging populations, unemployment, and stagnating and/or declining world economics. All of these changes have affected market dynamics and subsequently the strategic planning of the companies that operate in those markets.

Once a year, or whenever significant changes occur, strategic plans must be re-evaluated. With respect to the organization's quality efforts, the following issues must be addressed in the strategic plan:

- The rationale for any and all quality efforts.
- The philosophy and practices to be followed.
- The organizational values.
- The framework for managing the quality transformation(s).
- The infrastructure for supporting the quality initiative(s) and the integration of this support into policies, procedures, systems, and other associated strategies.
- A detailed operational plan for the integration of quality-related actions into current planning efforts.

7.3 The Information and Analysis Dimension

Managers make decisions. These decisions should be based upon achieving the strategic goals and objectives of the organization. The best decisions are those based on the analysis of data and information.

The biggest cost driver in any business is uncertainty, which is generally caused by a lack of accurate information about current and future trends and events. This uncertainty coupled with the decision maker's level of risk aversions results in costly safeguard actions such as safety stocks of inventory or in a high percentage of bad decisions. The solution to this problem is the acquisition of timely and accurate information followed by good analysis.

Deming pointed out that effective decision making is based upon facts, which in turn requires the collection and analysis of information on customer needs and expectations, operational problems, and the success of improvement projects. Numerous studies have observed that organizations that consistently collect and analyze data concerning their customers, their processes, and their improvement activities are more successful than those that do not. Organizational theory also argues that information processing is strongly related to organizational effectiveness.

There are several reasons behind why an analytical approach to decision making is important. The first is that with all of our connectivity, both internally and externally, there is an overwhelming amount of data that the average decision maker has to deal with. This increase in data volume causes what is called "data overload"; you cannot see what is important because of all the noise. Fortunately, with powerful and inexpensive computers and software, data analysis is becoming easier. Second, the complexity associated with the need to differentiate products based upon offerings, optimal prices, and inventories, and the endless number of interactions associated with these decisions is making it harder to make choices that lead to profits and increased competitive advantage.

As competitive pressures shorten the time to market for new products and the time to resolve customer issues, it is imperative that both information and responses are timely. The ability to make these types of decisions requires three things: a data collection process, a data management system, and analysis tools.

Data are a quantitative-based value assessment of a given situation, action, or result. For managers to make effective decisions, they need to understand where they are with respect to where they were in the past and where their competitors are now as well as in the past. They also need to understand what customer wants and how those desires have changed over time. To gain these insights, measurements need to be made and the data collected. For data to be useful, decision makers must remember that measurements are only assessments that we make. Thus, results are only meaningful in the context in which the data were collected, including the rationale behind why it was collected and the identity of the person(s) actually collecting the information it. Thus, data should be collected for a specific purpose and used only for analyzing that particular situation.

The decision analysis process is a logic-based approach for balancing the factors that influence a decision. This process often incorporates the uncertainty, values, and preferences of the decision maker as well as the technical, marketing, and environmental factors defining the problem. The primary objective in the process is to transform the plethora of disorganized and disperse data that organization collect overtime into high-quality, value-adding information that can enable individuals to make more effective decisions. A second objective is to analyze and interpret this complex technical information so as to make it coherent for presentation to non-specialists and to gain insights into root causes of problems or trends in process, organizational, and/or market behaviors.

For example, managers want to maximize their profits by having sufficient products to sell to customers. Thus, it is important to know how much inventory is needed. If they carry too much inventory, their expenses increase. If they carry too little inventory, they not only lose sales, but they could also lose customers who might not desert them for other suppliers. This will not only lower revenues but also increase expenses in trying to replace these lost customers. By collecting information on actual historical sales data for the various product offerings, a company can determine what the average revenues are for a given time period, as well as the standard deviation about these sales. The analysis can often tell the manager what the trend in sales actually is and whether or not there are any seasonal effects in the market. This information about sales characteristics can then be applied in inventory management models to determine an appropriate level of inventory to maximize selling opportunities while minimizing selling expenses. This same logic can be applied to virtually every situation that management faces.

Decisions are basically choices from among alternative courses of actions that yield uncertain futures, for which we have preferences. The three pillars of the

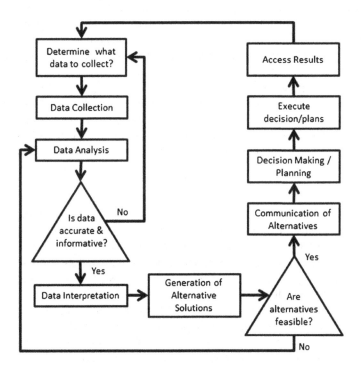

Fig. 7.1 The scientific methodology for decision making

decision-making process are as follows: what do you know; what can you do; and what do you want? The "what do you know" element focuses upon information you have. The "what can you do" element analyzes the information to generate alternatives. The "what do you want" element centers on your preferred results.

The quality of any decision is a function of the framework for the decision, the creativity of the alternatives, reliable information, preferences of possible futures, the soundness of the logic applied, and the commitment level toward solving the problem. There are many philosophies and methodologies for dealing with operational and quality problems. The six sigma methodology is a widely accepted and proven approach. Figure 7.1 illustrates the general methodology for the collection and analysis of data and decision making.

7.4 The Human Resource Management Dimension

The reason that entrepreneurs hire people to work for them is because there is simply too much work to be done for one person to create and deliver products and services to marketplace large than one. Having a diverse workforce with a breadth of skills and knowledge is essential for any organization to be success in today's marketplace.

Though the TQM philosophy is universally applicable, success of the program is highly dependent upon the organization's cultural fit with the tenets of the philosophy. Thus, management's focus must be on the integration between organizational structure and culture, management style, the workforce's TQM experience, and the market environment.

There are two principal aspects to quality: quality design and conformance to specifications. There are two paramount assumptions associated with the concept of a quality design: first, all of the customer's requirements have been captured and addressed in the design, and second, that a set of specifications based upon the design will yield repeatable results that will satisfy the customer. The second aspect relies heavily on there being a stable transformation process capable of producing products (or delivering services) that conform to the specifications of the design.

Where traditionally management of quality was concerned with the performance of the production process, TQM stresses the empowerment of employees to encourage both the improvement of a moral code and their sense of ownership toward meeting the requirements of internal customers as well as external customers. With TQM, employees are part of the delivery system. The emphasis of the TQM approach is on all employees being involved and responsible for serving the consumer. With this level of employee participation, quality matters throughout the transformation process from first supplier to delivery. Inspections are no longer the purview of a specialized department, but are now part of the job responsibilities of every employee. For this degree of employee involvement and interaction, teamwork and cooperation are essential.

The objective of TQM is to make every employee accountable for the quality of their workmanship and to get them committed to attaining quality. In return, management must assume total responsibility for ensuring there is a quality-focused infrastructure in place to facilitate that process and assure the results. Any deviations or lack of consistence in policies toward quality can, and often do, lead to declines on the firm's competitiveness.

Taguchi and Crosby believed that quality awareness was created by communication of the obstacles employees would face in their jobs. Deming, Feigenbaum, and others advocated all employees need to be involved in the decision-making process. Ultimately, quality results rely upon trained and committed employees. People are the source of ideas and innovation; people possess expertise, experience, and knowledge, and, only people can integrate these assets to generate products and services that costumers will value.

In the TQM philosophy, employee involvement is defined as employee participation in decision making and problem-solving process. Due to the increased pressure on employee associated with the expansion of their traditional responsibilities, there needs to be an organizational culture in place that when mistakes are made, the default question is what happened and how can we fix it so it does not happen again not who is to blame? Where workers possess firm-specific knowledge, companies have been able to improve their competitive positions.

Dr. Guest, Professor in Organizational Psychology and Human Resource Management at Kings College in London, stated that "because they are the most variable and the least easy to understand and control of all management resources, effective utilization of human resources is likely to give organizations a significant competitive advantage. The human resource dimension must therefore be fully integrated into the strategic planning process" [1].

7.5 The Process Management Dimension

In our dynamic world, staying static is to die. If you are producing a product or delivering a service and are making a sustainable profit, you will soon have competition for that market share. The more profits you make, the greater the amount of competition. Furthermore, the only way your competition is going to win your market share from you is to be better than you are. Thus, continual improvement is paramount in order to survive.

Due to the increased competitive pressure from globalization, companies can no longer sustain their competitive advances with a product-oriented philosophy. Instead, management has had to adjust their perspectives toward a customer orientation. This new orientation requires new management concepts and faster reaction times to changing market demands. In this new environment, the quality of production and business process becomes a central focus of management. With this new focus, strict organizational structures are being replaced with a process orientation where the identification and management of cross-functional processes is the key to success.

Process management is a best practice-oriented management principle focused on the sustainability of the firm's competitive advantage. The goal of process management is to breakdown all of the activities within a company into processes and align them with strategic goals.

When developing or improving a process, Porter points out that if the "best practice" is equality beneficial to all those seeking to adopt it, then the practice will not yield a sustainable competitive benefit. The key to conferring lasting benefits is to link the process design with firm-specific routines and capabilities. Thus, process management practices should focus on increasing the "fit" between those "firm-specific" routines and capabilities and complimentary activities across the organization.

When new customer needs are recognized and after products and/or services are designed, processes must be developed to produce them. The objective of development activity is to design a process capable of producing a desired result efficiently. The typical steps in this development process are as follows:

- Define the process for achieving the desired result.
- Identify and the critical to quality metrics associated with the inputs and outputs of the process.
- Identify the interfaces between the process and the various functions of the organization.
- Identify the internal and external customers, suppliers, and other stakeholders of the process.
- Evaluate possible risks, consequences, and impacts of the process on customers, suppliers, and other stakeholders.
- Establish clear lines of responsibility, authority, and accountability for managing the process.
- Document the process steps, activities, flows, control measures, training requirements, equipment, methods, information, materials, and other resources needed to operate and maintain the process.

7.6 The Innovation Dimension

The foundation of innovation is creativity. Innovation is the successful implementation of creative ideas. Given that level of competition, and the increasing rate of change in the marketplace, innovation is vital to the long-term success of every company.

Creativity is defined as the production of novel and appropriate ideas to everyday life. Research has found that people are at their most creative with motivated by intrinsic factors as opposed to extrinsic factors. Intrinsic motivation drives us to work at a task because we find the task to be interest, exciting, satisfying, or personally challenging. Extrinsic motivating factors include expected evaluations, surveillance, competition, dictates from superiors, or the promise of rewards.

Though intrinsic motivation is in large part inherent to an individual's person-ality, one's social environment can have a significant influence on the degree to which these motivating factors affect creativity. For example, Einstein said that his creativity was greatly dampened by the coercive environment of a militaristic classroom setting at the Luitpold Gymnasium.

The componential theory of creativity states that everyone has the capabilities to produce creative work in some domain, some of the time, and that social envi-ronment can influence both the level and the frequency of that creative behavior. This theory holds that there are three major components to an individual's creativ-ity: expertise, creative thinking skill, and intrinsic task motivation. At the intersec-tion of these three characteristics, creativity exists.

The two skill components (expertise and creative thinking) determine a per-son's capability for doing work in a given domain. Expertise is the cognitive pathway given to problem solving. These pathways include memory for factual knowledge, technical proficiency, and special talents toward the work within the target domain. Creative thinking skills include a cognitive style favorable to taking new perspectives on problems, the application of techniques for the exploration of new ideas, and a work style conducive to persistence. Personal characteristics, such as independence, self-discipline, risk-taking, tolerance for ambiguity, perse-verance, and a relative lack of concern for social acceptance, are key factors in the make up of one's creative thinking abilities. The task motivation component determines what a person will actually do. Though motivating factors are either intrinsic or extrinsic, intrinsic factors are more conducive to creativity than extrin-sic motivators.

It is commonly assumed that the relationship between intrinsic and extrin-sic motivation is antagonistic. In other words, if extrinsic motivation for a given activity increases, then intrinsic motivation must decrease. In fact, there are condi-tions where extrinsic and intrinsic motivation can combine synergistically, posi-tively affecting creativity. There are three determinants to whether extrinsic factors combine positively or negatively with intrinsic factors: a person's initial motiva-tional state, the type of extrinsic motivation used, and the timing of the extrinsic motivation.

A person's initial state of motivation is affected by his/her attitude and motives. The more vague and ambiguous one's attitude and motives, the more he/she can be positively affected by external influences. Alternatively, if his/her intrinsic motiva-tion is relatively week, extrinsic motivators could negatively affect creativity.

When an extrinsic motivator either confirms the individual's competence at a task, or provides information important to the performance of the task, it is called an informational extrinsic motivator. When the extrinsic motivator directly increases an individual's involvement in a track, it is called an enabling extrin-sic motivator. Both informational and enabling extrinsic motivators can stimulate positive outcomes. On the other hand, controlling extrinsic motivators place con-straints on how a task can be performed and thus will be detrimental to intrinsic motivations and performance.

7.6.1 Developing Organizational Creativity

The creativity of employees and their respective work groups are the primary source of innovation within organizations. The most important driver of creative thinking comes from the work environment established within the company. The greatest impact of work environment is its influence on the individual task motivation. The overall motivation toward innovation is made up of the organization's cultural orientation to innovation and support of creative activities. This cultural orientation is a primary responsibility of upper management. The most important element of an innovative orientation is valuing creativity and innovation in general, an acceptance of reasonable risks, employee pride and enthusiasm toward accomplishment, and strategic goals and objectives that drive innovation.

If top management wishes to change the direction and/or focus of their organization, they must take a visible and active leadership role in the change throughout the duration of process and even beyond it. The primary reasons driving this role are as follows: management controls all the organization's resources, management aim to achieve the goals and objectives of the organization, and management establishes the recognition and reward structures for the organization. If management redirects the application of resources to a new set of goals and objectives, and the recognition and rewards are not aligned with the stated goals and objectives, the workforce will interpret the conflict as to meaning the management was not serious and did not actually intend for the change to occur.

Mechanisms that promote organization-wide support for innovations are as follows: the open and active communication of information and ideas; recognition and reward system for creative work; and a fair evaluation of work. The characteristics of successful innovation projects are as follows: earmarking of special funds for the project; a formal review process of the idea followed feasibility studies; consideration of market issues during the planning stage; flexible planning that allows for the modification of the original idea as needed to fix organizational and market constraints; a monitoring and control process; and prototyping and market testing of new ideas before full-scale implementation. Conversely, elements that will undermine creativity are as follows: political problems (e.g., turf wars); destructive criticism and competition within the company; micromanagement by superiors; and excessive levels of structural formality and procedures. Practices that help promote creativity and innovation are as follows:

- Education of the types of motivation, their sources, effects on performance, and susceptibility to environmental influences.
- Knowledge of the types of intrinsic and extrinsic motivators and the context in which they are presented.
- Aligning employees to tasks that effectively utilize their skills and are of valued to the organization.
- Management commitment toward innovation.
- Organizational-wide processes that promote innovation.

- The creation of cross-function work groups.
- The effective allocation of resources supporting employees in the performance of assigned tasks.

7.6.2 The Work Environment

There are five environmental components that affect creativity in the workplace: a general encouragement of creativity; autonomy in the day-to-day conduct of work; resources; workload pressures and creative challenges; and organizational impediments.

7.6.2.1 The Effects of Downsizing on Creativity

Periodically, organizations find themselves in a situation where they need to downsize in order to remain competitive. The primary motivation for downsizing is to realign organizational resources to increase productivity and/or to reduce expenses to market conditions. This is an intentional management action that involves a reduction in force. It has been argued that this action will generally have positive effects on a company's efficiency by reducing waste and leading to a more productive allocation of resources. Unfortunately, innovation and productivity often fall casualty to these types of events. The principle reason for this is that during downsizing periods, all of the environmental stimulants to creativity within the company decline, but typically gain in strength as downsizing drew to a close.

There are some common patterns of change to the work environment that organizations incur when downsizing:

1. Communications deterioration at several organizational levels leading to the perception of a lower emphasis on creativity;
2. Centralization of control which leads to a perception of lower autonomy, which in turn leads to a deterioration of trust;
3. Conservative allocation of resources leading to a perception of insufficient resources, causing a resistance to change and a tendency toward rigid behaviors; and
4. High level of uncertainty and chaos causing employees to be more reliant upon familiar routines and greater conservatism.

Research has found that the main creativity stimulating areas affected by downsizing are as follows: challenging opportunities, work group support, and organizational encouragement. Furthermore, organizational impediments were found to significantly increase during downsizing events. These impediments will negatively impacting creativity and productivity. Research has found that 4 months after the downsizing event(s) ended, only productivity rebounded to any significant degree.

Another interesting finding regarding the causes of creativity and productivity declines during downsizing is employee perceptions. The anticipation of downsizing has a strong negative affect on employee perceptions and behavior. The longer the expected downsizing period, the lower the morale and the less creative people are in their work. The strongest effects occur when the perceived impacts of the downsizing affect the stability of individual employees' work groups. Overall, because of the devastating effects on surviving employees' motivation and creativity, when companies face downsizing issues, they should make every effort to ensure that they go as deeply as they will need to on the first pass and to fully inform employees of the plan and its economy drivers. Anecdotal evidence suggests that most companies underestimate the extent of future downsizing that the company needs to go through. This optimism frequently leads to multiple downsizing events, which in turn leads to exponential losses in morale and creativity across the organization.

7.7 The Supplier Relationship Management Dimension

Supplier networks are following the often quoted dictum of operating computers: "Garbage in equal's garbage out." One of Deming's 14 points for management was to move toward a single supplier system, where there was only one supplier for any given items. The driving logic for this charge is that multiple suppliers frequently mean variation between feedstocks.

Suppliers come in two varieties: external and internal. The external supplier provides the raw materials, components, and services necessary for the organization to create and deliver its products and services to its customers. The internal supplier is a member of the organization and provides the inputs necessary for other members of the organization to effectively perform the tasks they are responsible for. In both cases, it is essential that the supplier provides the customer with consistent, high-quality input in order to achieve the planned level of quality outputs needed for the successful creation of value.

7.7.1 The Supplier Relationship Management Process

Supplier relationship management (SRM) provides a systemic process for assessing suppliers' assets and capabilities with respect to their ability to the firm's strategic goals and objectives. The process determines which activities to engage in with a given supplier. It also includes plans and execution interactions with the supplier in a coordinated manner during the life of the relationship. The goal is to maximize the value realized in the relationship.

The objective of SRM is to cultivate collaborative relationships with key suppliers to foster greater coordination and synchronization in the flow of both materials and information. The processes that facilitate flow of information from the consumer back up the supply chain and the flow of products/services/materials down the supply chain are as follows: SRM, customer relationship management (CRM), demand management, order management, manufacturing flow management, product development and commercialization, and returns management. When these cross-organizational processes are executed effectively, customers benefit through improvement of competitive capabilities, greater production efficiency, and improved product effectiveness.

7.8 The Customer Relationship Management Dimension

Without customers, there are not markets to service. Thus, there should be no need for companies to exist. Deming stated that, "the consumer is the most important part of the production line. Quality should be aimed at the needs of the consumer, present and future" [2]. In achieving this goal, companies need to strike a balance between consumers and their other stakeholders (e.g., investors, employees, suppliers, and society at large).

Quality is largely a customer perception that is often based upon how a company's products or service meet the customers' respective needs and expectations. TQM research findings suggest that by focusing upon increasing customer satisfaction, the firm's performance should experience significant and positive benefits.

The best method for increasing customer satisfaction levels is by gaining a better understanding of their needs and expectations, so as to develop products and services that fulfill them. In fact, the benefits from TQM practices increase as markets become more competitive, and the key practice in this philosophy gains organization-wide customer focus.

7.8.1 Customer Relationship Management Process

CRM practices provide the structure for how companies develop and maintain relationships with their customers. These relationships go beyond just the consumer to include those companies in the supply chain between your company and the final consumer of the product.

CRM is a micro-level process that when performed correctly can significantly and positively affect profit growth, customer satisfaction and loyalty, and the value of goods and services delivered to the consumer. Both SRM and CRM form the linkages that hold supply chains together and allow them to function efficiently. The goal of CRM is to increase the joint profitability of both customer and supplier by improving their respective abilities to create value.

For CRM to work effectively, it must be treated as a strategic issue that is cross-functional in nature and process based. The principle reason for this view is that knowledge is the fundamental source of the firm's competitive advantage. Even though skill-based knowledge is very important for the development and creation of products and services, knowledge of the customer is needed in order to effectively direct the application of skill-based knowledge.

The CRM process is composed of two processes: the strategic process, which establishes and provides strategic direction for the process, and the operational process where implementation occurs. In the strategic process, executive management will identify which customers are crucial to the company's success and determine how the relationship with those customer groups is to be developed and maintained. During the strategic process, executives will establish the criteria for the segmenting of customer within target markets and provide the guidelines for qualifying customers for product and service agreements (PSAs). These criteria often include metrics for profitability, growth potential, market access, market share, profit margins, technology levels, and buying behaviors.

At the operational level, customer segments are identified and accounts set up. A customer team is put together for each segment and tasked with increasing customer loyalty through the customization of products and services. The teams will themselves develop guidelines for the differentiation of the PSAs, including alternatives that consider revenue and cost implications. These teams will contact customers in their respective market segments that meet the criteria set by the company's strategic policies and craft PSAs that will meet their individual needs. The goal of these PSAs is to increase the profitability of both the firm and its customers. The achievement of this goal is enhanced through the development of long-term relationships between the firm, its suppliers, and its customers as depicted in Fig. 7.2.

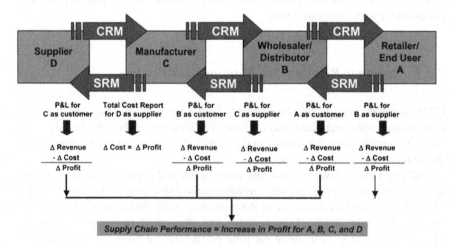

Fig. 7.2 Supply chain relationships (Published with kind permission of © Emerald Group Publishing, per the STM permissions guidelines. *Source* Lambert and Pohlen [3]. All rights reserved)

7.9 The Organizational Culture Dimension

An organization's culture is in effect an amalgam of those commonly held norms, values, beliefs, and attitudes of its members. Philosophers have posed several definitions of organizational culture:

- Organizational culture is the shared philosophy, ideology, values, assumptions, beliefs, hopes, behaviors, and norms that crate unity across the organization [4].
- The "common perceptions which are held by the members of an organization: a system of common meaning" [5].
- The "informal design of values, norms that control the way people and groups within the organization interact through each other and with parties outside the organization" [6].

According to the competing value framework, there are four principal types of organizational cultures and six dimensions to the cultures. In this framework, there are implications associated with each cultural type as follows:

- *The Clan Culture*: This cultural type is full of shared values and common goals, with an atmosphere of collectivity and mutual assistance, which stresses empowerment and member growth. This cultural type has a strong sense of family, where teamwork is emphasized and the leaders are mentors. Loyalty, cohesiveness, and participation are key components of success. The overall organizational focus is on the maintenance of stability.
- *The Adhocracy Culture*: This cultural type gives its members tremendous latitude to grow on their own as long as their actions are consistent with organizational goals. The leaders within the culture are characterized as entrepreneurs who are driven by innovation and finding new ideas. The overall organizational focus is on gaining market share and maximizing external developing opportunities. Individuals are recognized and rewards for creativity and innovation.
- *The Market Culture*: This cultural type is outwardly focused (e.g., on market transactions). Its main goal is to maximize profits through market competition. This culture stresses effectiveness in achieving the goal. Internal competition between members is common and is perceived as a development tool, and success is based upon individual achievement.
- *The Hierarchy Culture*: This cultural type operates on a clear structure, standardized rules and procedures, strict control, and well-defined responsibilities. Internal stability is paramount and is maintained through adherence to set of rules. Success is measured by task accomplishment in accordance with procedures and avoidance of destabilizing activities.

Organizations will always experience some amount of tensions between the various effectiveness attributes and their dominant cultural type. For example, organizations need stability, while at the same time, they need flexibility to address changing market demands. Thus, they need to maintain control and disciple

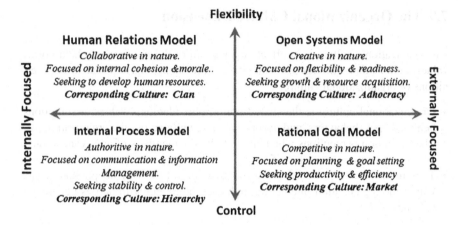

Fig. 7.3 The competing value framework for organizational effectiveness (Published with kind permission of © Sage Publications. *Source* Quinn et al. [7]. All rights reserved)

within the ranks, while still allowing employees freedom to innovate. The degree of effectiveness that can be achieved by these types of organizational cultures is described in Fig. 7.3.

Research has found that organizational culture has a direct effect on the performance of the firm. For example, cultures that value competitiveness and innovation will outperformance other cultures in dynamic and fluid market conditions.

7.10 Implications for the Customer

Customer satisfaction is achieved when the perceived performance and benefits of a product/service meet or exceed their original expectations. These expectations are often based upon the customer's credence about the offering, as modified by advertising and other promotioning materials and the perceived value of the offering. If the product/service exceeds or meets the original expectations, the customer's perceptions result in a positive opinion of the company's brand. The opposite is likely to happen if expectations are not met. The collective associations with the value from the company's product offerings result in brand equity. Positive brand equity is based essentially on the company targeting the product offerings at the correct quality points, so that consumers gain value from its usage. This accumulated strength can help the company on other products as well and therefore has carry over effects. However, at the same time, products/services that disappoint customers in their perceived quality might negatively other offerings in the company's product line.

The first step in improving an organization's quality focus is to recognize that the company is a collection of people, who execute a series of interconnected, interdependent processes in order to achieve specific goals and objectives. Next is to understand that the continued superior performance of the organization is predicated on the continuous improvement of both these processes and the knowledge and skills of the workforce. Frequently, it has been found that even though knowledge was available, resources were properly staged, and tools, processes, and work methods prepared that if the people implementing the solution were not working together to execute plans, nothing will occur. The key to remember about success is that "ITS ALL ABOUT THE PEOPLE."

TQM is a people-focused management system whose goal is to continual increase customer satisfaction. The foundation of the philosophy is the "scientific method." TQM includes not only people, but also systems, methods, and tools. Systems provide the structure within which to execute the philosophy, and the values espoused within the philosophy stress the dignity of the individual and the power of community action (e.g., teamwork). Even though people, systems, methods, and tools may change over time, the philosophy and values of TQM provide the continuity that the organization needs to remain competitive.

The success of the firm in achieving its quality goals is perceived and measured along the following dimensions: strategic planning, customer focus and satisfaction, human resources development and management, information and analysis, process quality management, innovativeness, behavior (e.g., organizational structure and culture), and SRM. The extents to which the various customer groups have visibility of these dimensions are noted in Fig. 7.4. Note that

Dimensions of Organizational Quality	Stakeholders			
	Investors	Employees	Customers / Consumers	Society
Leadership (Top Management Commitment)	✓	✓		
Strategic Planning	?	✓		
Customer Focus & Satisfaction	✓	✓	✓	
HR Dev. & Mgmt. (Employee Engagement)		✓		?
Information & Analysis (Problem Solving)		✓		
Process Orientation	?	✓		
Innovativeness (Continuous Improvement and Problem Solving)	✓	✓	✓	?
Behavior (Organizational Structure & Culture)	✓	✓	✓	?
Supplier Mgmt.	?	✓		

Fig. 7.4 Visibility of the dimensions of organizational quality to customers

the symbol "E" indicates total visibility to all customers and that the symbol "?" indicated only a few customers have visibility of that dimension.

Though employees and many of the firm's commercial customers have visibility and/or actual experience with its inner workings, many of its investors and virtually all of the consumers of their products and service have none. This latter group of stakeholders forms their perceptions of the company based upon credence and search factors. These perceptions often have a significant influence upon whether or not they will even look at, much less experience the company's market offerings.

References

1. Guest, D. E. (1987). Human resource management and industrial relations. *Journal of Management Studies, 24*(5), 503–521.
2. Deming, W. E. (1982). *Out of the crisis.* Cambridge, MA: Massachusetts Institute of Technology, Center for Advanced Engineering Studies.
3. Lambert, D. M., & Pohlen, T. L. (2001). Supply chain metrics. *International Journal of Logistics Management, 12*(1), 14.
4. Kilmann, R. H. (1985). Corporate culture: Managing the intangible style of corporate life may be the key to avoiding stagnation. *Psychology Today, 19*(4), 62–68.
5. Robbins, S. P., & Judge, T. A. (1984). *Essentials of organizational behavior.* Upper Saddle River, NJ: Prentice Hall.
6. George, J. M., & Jones, G. R. (2002). *Organizational behavior.* Upper Saddle River, NJ: Prentice Hall.
7. Quinn, R. E., Hildebrandt, H. W., Rogers, P. S., & Thompson, M. P. (1991). A competing values framework for analyzing presentational communications in management contexts. *Journal of Business Communications, 28*(3), 213–232.

Chapter 8
Implementing Organizational Change

The marketplace is a constantly evolving, highly competitive business environment. This environment is affected by many factors: changes in social norms; by government actions; by shifting demographics; by innovations in technology; by competitive actions of other business entities; by changing weather patterns, and by many other factors. Any change in the business environment creates both problems and opportunities for commercial organizations. In order to deal with these changes, companies will organize themselves in various ways that management deems advantageous to their goals and objectives. When the marketplace changes, the ability of these organizational alignments to create competitive advantages is either enhanced or impeded. Thus, in order to stay competitive, organizations must evolve to meet these new demands.

The process of adapting an organization to environmental changes and uncertainty is both time consuming and enormously complex. The myriad decisions involved in this process often encompass behaviors at several levels within the organization. The process of either designing a new organization or evolving an existing one involves finding solutions to three broad problems: the entrepreneurial problem, the engineering problem, and the administrative problem.

In new companies, the entrepreneurial problem is solved by developing a concrete definition of the domain that the company is competing in with specific products and/or services, and their respective target markets or market segments. With existing companies, there are the added dimensions of how to change or modify current products and/or markets within the constraints of their existing processes and the development of the necessary (i.e., new) market orientation.

The engineering problem is solved with the development of systems and processes which support the company's entrepreneurial solution. This development process involves the selection of an appropriate technology for the production and

© Springer-Verlag London 2015
G.N. Kenyon and K.C. Sen, *The Perception of Quality*,
DOI 10.1007/978-1-4471-6627-6_8

distribution of the new or modified products or services and the retraining of the organization to the new technology.

As part of the implementation of the engineering solution, management must address the administrative problem: strategies, structure, culture, and the reduction of uncertainty.

8.1 Organizational Improvement Methodologies

Organization development (OD) is a planned, organization-wide, collaborative effort between top-management and employees that is focused on increasing the organization's effectiveness, efficiency, and productivity. It has been described as a systemic learning and development strategy for changing the basics of beliefs, attitudes and values, and structure of the organization so as to better absorb disruptive technologies, changing market opportunities and ensuing challenges and chaos.

This OD process is executed through a series of strategic interventions to the organization's "processes," using knowledge that is based on behavioral science. Improvements are achieved by increasing the alignment of the business processes with the organization's competitive strategies, as well as dealing with the sociotechnical aspects of change. The interventions include methodologies and approaches to strategic planning, organization design, leadership development, change management, performance management, coaching, diversity, team building, and work/life balance. The principal objective of OD is to improve organizational performance while

- Increasing interpersonal trust between employees.
- Increasing employee satisfaction and commitment.
- Addressing problems and improving problem-solving skills.
- Effectively managing conflict.
- Increasing the level of cooperation and collaboration between employees.
- Increasing employees' awareness and respect for each other.

As a field of science, OD is a process centered on the continuous diagnosis, action planning, implementation, and evaluation of an organization's self-management capabilities. In addition, it encourages employees to improve their knowledge and skills in specific areas so as to improve the company's overall capacity for solving problems and managing future change.

The foundational theories of OD are based upon the realization that organizational structures and processes have a significant influence on worker behavior and motivation. OD is also focused on aligning organizations with their rapidly changing and complex environments by means of organizational learning, knowledge management, and the transformation of organizational norms and values. In creating, changing, and managing an organization's development, there are several key elements:

8.1.1 Organizational Climate

With any discussion of organizational climate, four basic assumptions need to be understood:

- Organizations operating in similar circumstances can react differently as well as perform differently with respect to their productivity.
- Productivity is related to individual as well as group motivations.
- Motivation will be affected, either positively or negatively, by the climate in the organization.
- Management shapes the organizational climate.

Organizational climate has been defined in several ways; to list a few:

- It is intuitively sensed rather than something that is cognitively recognized.
- It is a set of attributes which can be perceived about the organization or induced from the way the organization deals with its employees and environment.
- It is the collective view of employees as to the nature of the work environment in the organization.

Simply put, organizational climate is the set of properties that characterize the working environment, which are perceived either directly or indirectly by employees, and can greatly affect their behavior. Employee attitudes and beliefs regarding company practices create the organization's climate. The key factors determining an organization's climate are as follows: leadership, management style, communication, organizational structure, the recognition and rewards system, and organizational goals.

8.1.2 Organizational Culture

Culture is defined as the deeply seated norms, values, and behaviors that an organization's members share. The five basic elements that form the basis of an organization's culture are as follows: assumptions, values, behavioral norms, behavioral patterns, and artifacts. The elements of assumptions, values, and norms (i.e., subjective features) reflect the members' unconscious thoughts and interpretations of their organizations. These subjective features in turn influence the behaviors of the members. Artifacts provide the physical representations of members' collective assumptions, values, and norms; thus, providing reinforcement to the desired behaviors.

8.1.3 Organizational Structure

Structure defines how power and responsibility flow through an organization. The goal of any given structure is the efficient and effective execution of organizational plans and strategies for the achieving of objectives, through the activities of task allocation, coordination, and supervision. Organizational structure affects the achievement of objectives in two ways: (1) by providing the foundation on which standard operating procedures and routines rest and (2) by allocating selected sets of individuals to specific decision-making processes.

There are several structure types and three basic characteristics. The type chosen by an organization should be adaptive to process requirements. This will optimize the ratio of effort and input to output.

8.1.3.1 The Pre-bureaucratic Structure

The pre-bureaucratic structure is best suited for organizations that do not have a significant number of standardized tasks. This is typical of many entrepreneurial companies. This type of structure is totally centralized. The principle of the company typically makes all of the key decisions and most communications is performed one on one. This structure allows the principle to the control growth and development of the company.

8.1.3.2 The Bureaucratic Structure

When a company grows beyond the point where a single individual can effectively manage and control its processes, the bureaucratic structure offers greater control and efficiency. This type of organizational structure is characterized by well-defined roles and responsibilities, a hierarchical reporting structure, and an institutional respect for the merit-based recognition and reward system.

As opposed to the pre-bureaucratic structure, the bureaucratic organizational structure has a certain degree of standardization; thus, it is better suited to complex or large scale operations. This structure relies heavily on rigid and tight procedures, policies, and constraints. Unfortunately, this tendency also makes the organization reluctant to change and harder to adapt to shifts in the marketplace or business environment. Other disadvantages associated with this type of structure are that they can discourage creativity and innovation.

8.1.3.3 The Post-bureaucratic Structure

In the post-bureaucratic organizational structure, there is a focus on moving past the traditional inflexibilities of the pure bureaucratic structure. This type of structure can be created in two ways. In the generic sense, new management philosophies such as total quality management, cultural management, matrix management, or others are adapted. The organization still has its core tenets of bureaucracy. Its hierarchies still continue to exist. The organization will still be rule bound, and authority will still be based on a rational, legally based criterion.

The other method for transitioning for a bureaucratic to a post-bureaucratic structure is to shift decision making from autocratic to consensus. In this type of culture change, the emphasis is on meta-decision-making rules. The objective is to encourage employee participation in the process and to empower them.

8.1.3.4 The Functional Structure

For companies producing standardized products and/or services at low costs for large markets, this organizational structure can improve organizational efficiencies. In this structure, employees are grouped according to the tasks they perform. This allows them to become specialists within their realms of expertise. Control and coordination emanate from the hierarchy within the functional group. The most common functions found within companies are as follows: production, marketing, finance and accounting, and human resources.

The disadvantages of this structure include:

- Communications between functional divisions can be rigid due to standardized operations and a high degree of formalization. This is likely to make the organization slow and inflexible.
- The level of cooperation between functional groups can be compromised by territorialism and infighting.

8.1.3.5 The Divisional Structure

When segregation of business activities is advantageous, the division structure can increase flexibility and efficiency. In this structure, the company will divide itself into operational groups called divisions. These divisions are self-contained business units, or profit centers, that consist of a collection of function which work to produce a product and/or service.

The advantage of a divisional structure is that it allows for the delegation of authority to each group. It also facilitates the measuring of that performance more directly. Furthermore, this structure allows divisions to focus and specialize on a given product, product line, market, and/or geographical region. Managers also perform better and employees tend to have higher levels of morale. In addition, the divisional structure is more effective in coordinating work between divisions, than a corresponding functional structure.

The disadvantages of the divisional structure are that it can support unhealthy rivalries between divisions. It can also increase costs and there can be an overemphasis on divisional goals at the expense of corporate goals.

8.1.3.6 The Matrix Structure

For organizations that constantly have a constant stream of multiple project-styled activities, the matrix structure is more dynamic than the function structure. The matrix structure combines all of the other structures and allows the project teams to share information more readily across task boundaries, while simultaneously allowing for specialization.

The disadvantages of the matrix structure are that it increases the complexity of the chain of command, and it increases the manager to worker ratio, which often result in conflicting employee loyalties.

8.1.3.7 The Flat Structure

It is the natural tendency of bureaucracies to grow overtime, adding layers of hierarchy to its administration. This tendency dramatically slows decision making which in turn reduces the organization's competitiveness. By increasing the horizontal breath of the chain of command and reducing its vertical hierarchy, companies can recapture much of the agility typically associated with smaller, entrepreneurial types of organizations.

8.1.3.8 Structural Characteristics

Three are three basic structural characteristics that significantly affect/influence organizational communications, coordination, and decision making: formalization, centralization, and specialization.

Formalization

Formalization is defined as the degree to which decisions and relationships are governed by formal rules and procedures. These rules and procedures in turn provide a means for defining appropriate behaviors. Routine aspects of a problem can be dealt with easily through the application of these rules, and the rules enable individuals to organize their activities to benefit themselves and the organization. These articles of formality are a form of organizational memory and enable businesses to fully exploit previous discoveries and innovations. Formal rule and procedures can also lead to increased efficiency and lower administrative costs.

Centralization

Centralization denotes whether or not decision authority is closely held with executive managers or if it is delegated down to middle and lower level managers. Lines of communication and responsibility are relatively clear in centralized organizations, and the route to executive management for approval can be travelled quickly. While fewer innovative ideas might be put forth in centralized organizations, implementation of these ideas tends to be straightforward once a decision is made. This principle benefit to centralization is realized in a stable non-complex environment.

Specialization

Specialization refers to the degree to which tasks and activities are divided in the organization and are assigned to work groups that specialize in the related tasks. Highly specialized organizations have a higher proportion of highly skilled personal called "specialists" who direct the effort in a well-defined set of activities. Specialists are experts in their respective areas and typically are given substantial autonomy, which enables the organization to respond rapidly to changes in its environment. Organizations that have a high proportion of generalists are typically low in knowledge about their area of focus.

8.1.4 Organizational Strategies

Organizational strategies are an articulation of how the organization needs to evolve overtime in order to meet long-term objectives. It is a framework for instituting the required changes needed to achieve those objectives, as well as a timeline for specific performance goals. The process of evolving an organization to meet its future needs and/or adjust to environmental changes is immensely complex and involves a wide range of decisions and behaviors at multiple levels of the organization.

In developing strategies for the business, management must first compare the company's current situation to a targeted future state; thus identifying the gaps between the two states. Then goals and objectives need to be defined based upon what is required for the desired changes to take place. To be effective, these strategies must be realistic and measurable, while at the same time, forcing the organization to stretch and grow its capabilities.

Numerous organizational gurus believe that the behaviors of an organization are only partially due to environmental conditions, with management decisions being the critical determinants of structure, process, and culture. Three broad problems have been identified in addressing organizational change: the entrepreneurial problem, the engineering problem, and the administrative problem.

8.1.4.1 The Entrepreneurial Problem

For new companies, the entrepreneurial problem faced in adapting an organization to effectively handle future market conditions involve how an idea that is vaguely defined is developed into an organizational domain (i.e., a specific good or service and a target market or market segment). For established organizations, the entrepreneurial problem is similar to that of the new company, except with an additional dimension: How to fit the new domain into an existing set of solutions for its "engineering" and "administrative" problems. The entrepreneurial energy that exists with its current domains is not easily transferred as a result of the stability that the organization has currently achieved. In both situations, the solution is marked by management's choice of a particular product–market domain. The solution to the entrepreneurial problem is usually sought through the development and projection of an organizational "image" that defines the product–market domain and the company's approach toward commercializing that market.

8.1.4.2 The Engineering Problem

The engineering problem in organizational change involves the operationalizing of management's solution to the entrepreneurial problem. The crux of this problem is the selection of an appropriate technology for the production and delivery of the chosen product and/or service. Solving this problem also includes the development of information, communication channels, and control linkages for insuring the proper operation of the chosen technology.

8.1.4.3 The Administrative Problem

The administrative problem concerns the organization's socio-technical systems. Socio-technical theory is founded on two main principles:

- The interaction of social factors and technical factors creates the environment for successful, or unsuccessful, organizational performance.
- Optimization of each aspect (e.g., social or technical) alone tends to increase not only the quantity of unpredictable relationships, but also those relationships that are detrimental to performance.

The goal in solving the administrative problem is to reduce the amount of uncertainty for employees in the organization. This effort often includes the rationalizing and stabilizing of activities used in solving organizational problems encountered while solving the entrepreneurial and engineering problems. The principal challenge here is to balance the social needs of the organization for stability and reduced uncertainty, with the organization's competitive needs for formulating and implementing processes which will enable it to continuously innovate and improve. The fact that the process is not linear further compounds the difficulty of solving these three problems. The three solutions must be derived and executed together, and frequently, there will be iterative course corrections before reaching an optimal state.

8.1.4.4 Strategic Topologies

It has been found that a company's overall performance is strongly influenced by the degree to which its business strategies are supported by its chosen organizational structure and culture. The key to the integration of these three components is management's ability to identify the key success factors, or activities, for their chosen strategy and developing the organizational structure and culture that best promotes those success-related activities. Marketing research suggests that there are four basic business strategies: prospectors, analyzers, low-cost defenders, and differentiated defenders.[1]

The Prospector Strategy

Prospectors seek to identify and exploit new product and market opportunities. In order to be successful in this endeavor, they must have core competencies in the areas of developing innovative new products and in market development. An example of a prospector is found in Apple during the Steve Jobs eras.

The Analyzer Strategy

Being slightly more risk adverse than prospectors, analyzers like to monitor customer reactions to both new products and competitor activities in the various markets. After noting and understanding the "whys" of the successes or failures of difference competitive activities or finding improvement opportunities with new products, analyzers will enter a market, seeking to either improve upon successful product offerings or offer comparable products at a lower price.

One of the major success stories in the computer manufacturing market is Dell. Dell was a "Johnny Come Lately" to computer manufacturing and marketing. While working on an IBM personal computer (PC), Michael Dell realized that the components that went into the construction of the computer could be purchased for one-fourth of its purchased price. Even factoring in the additional costs of increased memory, bigger monitors, and faster modems, a PC could still be assembled and sold for a handsome profit. The business implications of this knowledge were too great to ignore. Also, the time for low-cost alternative was just right; in the mid-1990s, the general public was becoming interested in computers.

Where the Big Three where using channel partners for sales and distribution of their products to customers, Dell created an online, direct to customer business model with at home support services. This approach allowed Dell to operate at a substantially low cost, while providing greater flexibility in product configurations

---

[1] Published with kind permission of © Elsevier Publications. *Source* Olson et al. [1]. All rights reserved.

and improved customer service. Today, Dell is still working on improving its business model by seeking out for new manufacturing and distribution models. Their goal is to give customers what they need and make the technology simpler and easier to use.

The Low-cost Defender Strategy

The Low-cost Defender primarily operates in stable market segments. Their overall emphasis is on efficiency through standardization rather than on effectiveness from flexibility. Examples of this type of strategy can be seen in airlines companies such as southwestern Airlines and Jet Blue.

The Differentiated Defender Strategy

In contrast to the low-cost strategy, differentiated defenders emphasis outstanding service and high quality, while trying to control their markets through superior products and service, and brand image. The Broadmoor Hotel and Resort is a good example of the differentiated defenders strategy.

8.2 Essential Elements for Change

One of the main impediments to any type of change is internal resistance. People by nature hate the unknown and change is the optimization of unknown. There are three essential elements to actually realizing strategic objectives of change: motivational leadership, action plans, and performance management.

8.2.1 Motivational Leadership

Leadership is all about the influencing of other peoples actions. The motivational aspect of leadership focuses on achieving sustainable performance through personal growth and planning that recognizes the dynamics of human nature. The core principles of motivational leadership are strong ethics, a clear vision of the goals and objectives for some future state of the organization, definable values, authentic communication, and being genuinely motivated to promote collaboration and positive energy throughout the company.

Motivating others involves the development of incentives, or conditions, that help move people toward a desired behavior. These incentives are either intrinsic or extrinsic and must be based upon the beliefs, values, personal interests, and even fears of the individuals being motivated.

8.2.2 Intrinsic Motivation

Ultimately, all motivation comes from within the individual. Even though there may be external factors influencing an individual's behavior, they play a secondary role in the motivational process. Before external factors can be effective as motivators, they must be in congruence with the individual's intrinsic factors.

In fact, many philosophers believe that there is only one kind of intrinsic motivation: Actions that enhance or maintain one's self-image. While others say that intrinsically motivating activities are those which are joined into for no other reward than the simple enjoyment that they bring. In support of the latter argument, there are seven ways of designing an organizational environment that are self-motivating.

8.2.2.1 Challenges

Confident- and achievement-oriented individuals are usually motivated by working on projects that have personally meaningful goals. Attainment of those goals needs to involve activities that are increasingly difficult, but achievable. Two of the best ways of designing this type of activities is to provide feedback loops that inform the individuals of their performance, and by aligning the goals of the tasks with the individual's self-esteem.

8.2.2.2 Curiosity

Individuals can often be motivated by activities that arouse their curiosity (i.e., stimulate their interest toward learning more). This type of work environment can be created by presenting individuals with work that connects with their present knowledge or skills at a more desirable level.

8.2.2.3 Control

We mentioned before that people hate the unknown. The fear is often associated with their lack of control i.e., the feeling of being adrift. By designing opportunities for employees to have some level of control in their work can increase motivation. In order to maintain this motivation, employees must understand the cause and effect relationship between the actions they take and the results they get. Methods for achieving this are to:

- Establish measurement and reward structures in the organization that make the cause and effect relationship clear.
- Develop an organizational climate and culture that allows individuals to believe that the work they perform does make a difference.
- Promote education and training systems where individuals can choose what they want to learn, and how to go about learning it.

8.2.2.4 Fantasy

Using fantasy as an intrinsically motivating factor is to use mental imagery of things and/or situations of possible futures that are related to the results of a given set of activities. For example, if an individual has aspirations of being in control, then talk to him about the future possibility of their being in charge of a larger and/or more important operation.

8.2.2.5 Competition

On many occasions, competition can be used to inspire and motivate individuals. This is because there is a level of satisfaction associated with comparing one's performance against that of others. This type of motivation technique can easily get out of control if not managed correctly. Thus, one should keep in mind the following:

- Competition is not universally appealing.
- Losing a competition will often demotivate more than winning motivates.
- If not kept in check, competitive spirits can reduce the level of cooperation in the workplace.

8.2.2.6 Cooperation

People are basically benevolent by nature, and by helping others to achieve their goals, the providers of the service also attain a degree of satisfaction. When crafting a cooperation culture, one should keep in mind:

- Cooperation is not universally important to all individuals.
- Cooperation is a valuable skill that can facilitate benefits in many different situations.
- Interpersonal skills are a critical factor for cooperation.

8.2.2.7 Recognition

Recognition is a great motivator. People are encouraged and feel a degree of accomplishment when their efforts are recognized others. One of the major success factors associated with recognition is that it cannot be based on comparisons of one worker's achievements to those of others. Instead, it should be basic upon measurable, sometimes incremental, achievements toward organizational goals and objectives.

8.2.3 Extrinsic Motivation

As previously mentioned, extrinsic or external motivations are factors outside of the individual's personality that stimulate their internal drive. The concept of externally motivating someone is not at odds with the fact that the actual drive comes from within them. Thus, it is possible to engineer situations, or an external environment, that fosters motivating feeling in an individual.

8.2.3.1 Employee Motivation

Recognition, providing positive performance feedback, and by challenging employees to learn new things are some of the most effective ways for managers and leaders to employees. Ineffective managers often make the mistake of introducing negative factors into the workplace that tend to demotivating employees, such as punishment for mistakes, or frequent criticisms.

In fostering improvements in employee motivation, management should design an environment that encourage employees to take higher degrees of responsibility and accountability for decisions associated with the tasks they are assigned. The key factor in controlling this type of system is to recognize that motivation is an individual trait. What works for one person will not necessarily work for the next, so there needs to be flexibility designed into the process that allows for varied degrees in applying this practice.

8.2.3.2 Figuring Out What Motivates Others

In an organization, people rely on their leaders to bring meaning to actions, thus helping them to find purpose and value in an environment that is demanding of results. Leadership is the common thread which runs through the entire process of translating planning into results. It is also the key to engaging the hearts and minds of the people involved with the process. Regardless of the intent, effective leadership often makes the difference between success and failure. The following tips can help in determining how to motivate others:

- Get to know the employees as individuals. Talk to them about what they value. This will provide insights into which of the seven factors mentioned above might be high on their list.
- Do not give up if the first attempt results in failure. Finding the correct motivation technique for an individual is often an iterative process. Test a factor on an employee: if it works great, if not try another.
- Maintain a running dialog with employees. Not only do you and a manager need to know what they are doing, but it helps in better understanding about

how they feel about different situations. It also allows you the opportunity to see how there are reacting to each factor. Getting feedback from employees is always a good idea.

- Always be on the guard for signs of demotivation. It is important to ensure that outside factors are not being introduced into the work environment causing employees to become counter-productive.

8.2.4 Action Planning

The best way to insure success is to plan for it. "Strategy into Action" planning is a phased approach charting a course through performance factors, linking strategic thrusts to project, and departmental and individual activity. The ultimate goal is to enable organizations to effectively translate strategic intent all the way through to results in a clear and powerful process.

The real need is to creatively and systematically unfold the strategy, bring it to life by creating integrated action plans across an organization that ensure all functions and divisions are aligned behind it. There are four basic steps to helping organizations negotiate change: Diagnosis, Action Planning, Intervention, and Evaluation.

8.2.4.1 Diagnosis

The diagnostic phase of changing an organization involves the identification of problems that may interfere with the effectiveness of the change and the strategy for implementing it, and/or the assessment of any underlying causes that may prevent a successful implementation. The process for identifying problems starts with examining the company's mission, goals, policies, structures and technologies—climate and culture; environmental factors; desired outcomes and readiness to take action.

The methods for performing this diagnosis have moved from being a purely behavioral analysis toward one of a strategic and holistic business diagnostic approach. Today, the diagnostic approach explores the interactions between management, employees, suppliers, and customers in the context in which they operate. The objective of the diagnosis is to better understand the company's strengths, deficiencies, and opportunities for improvement, and to later devise a targeted intervention and measurement strategy.

8.2.4.2 Action planning

Once a set of strategic interventions have been identified, specific actions need to be planned for executing them. The first step in this planning process is to assess the feasibility of implementing the identified change strategies.

8.2.4.3 Intervention

Change steps are specified and sequenced, progress monitored, and stakeholder commitment is cultivated.

8.2.4.4 Evaluation

Assess the planned change efforts by tracking the organization's progress in implementing the change and by documenting its impact on the organization.

8.2.5 Performance Management

Performance management involves the creation of organizational processes and capabilities necessary to achieve performance through people delivering results. With the changes to organizational strategies, management needs to review its policies and procedures to ensure that they support and promote the desired organizational behaviors. In other words, "YOU WILL GET THE RESULTS THAT YOU RECOGNIZE AND REWARD."

More often than not the best laid plans get derailed due to a failure to properly align incentives and controls with strategic goals and objectives. The best method for ensuring that everyone in the organization is engaged in executing the new course of action is to communicate the strategic intent, thrusts, and action plans, as well as, the benefits to everyone in the organization. The characteristics of an effective performance management system are as follows:

- An effective and comprehensive communicate strategy;
- A performance measure that is aligned with goals and objectives and captures and reports results in real time; and
- Acknowledge and enable emotional contracting with all employees in order to link organizational intent with the motivations, values, and aspirations of the people.

8.3 Driving the Change

Before you start trying to change anything about your organization, there are several questions that you must first ask yourself. What is it that you are specifically trying to achieve, and just how important is it? What is the value proposition for the consumer? What is the value proposition for my other customers? How can the change, and the benefits associated with the change, be made visible to customers? How can the change be measured during implementations? How can the effects of the change be measured and maintained overtime?

With these questions answered, you can then engage in defining the strategic goals and objectives of the change, developing action plans, and establishing roles and responsibilities for those groups and individuals that are involved with the change.

8.3.1 Tools for Managing Organizational Change

The first step in the improvement process is for the organization itself to understand how its delivery systems and key processes work. Both resources (i.e., inputs) and activities (i.e., processes) must be addressed in order to ensure that the quality improvements (i.e., outcomes and outputs) are achieved. This is not just a matter of developing or reviewing a flow chart and studying it; the organization's culture will have a tremendous effect on both the efficiency and the effectiveness of every process and system within the company.

8.3.1.1 FlowCharts

A flowchart is a diagram describing the flow of materials and information in a process. This description is illustrated by showing the steps in the process as boxes of various kinds, and their sequential ordering is denoted by connecting the boxes with arrows. Flowcharts are commonly used in analyzing, designing, documenting, and/or managing a process. The two most basic types of boxes used in a flowchart are given as follows:

- a processing step, called an *activity*, that is symbolized using a rectangular box,
- a decision that is symbolized using a diamond.

Depending its application, there are several common names for the flowchart, including: flowchart, process flowchart, functional flowchart, process map, process chart, functional process chart, business process model, process model, process flow diagram, work flow diagram, and business flow diagram. The typical flowchart symbols are as follows:

- Start/End events are represented by circles, ovals, or rounded (fillet) rectangles, usually containing the word "Start" or "End."
- Arrows are used to show the direction of the flow of control in the process. Thus, the control passes from the event where the arrow starts and ends with the event that the arrow points to.
- Generic processing events are represented as rectangular boxes.
- Subroutines are represented as rectangles with double-struck vertical edges. This symbol is typically used to show complex processing steps which may be detailed in a separate flowchart.
- Input/Output events are represented using a parallelogram.
- The preparation of conditional events is represented as a hexagon. This symbol represents an operation which has no effect on the process other than preparing a value for a subsequent conditional or decision step.

- Conditional or decision events are represented using a diamond which shows where a decision is necessary. The decision is typically a Yes/No question or True/False test. This symbol uses two outgoing arrows, each representing one of the conditional outcomes. When more than two arrows are used, indicating a complex decision, it may necessary to breakdown symbol further or replaced with the pre-defined process symbol.
- Junction symbol is generally used to represent where multiple control flows converge in a single exit flow. A junction symbol is typically denoted with a blackened circle and shows where more than one arrow is entering, while only one arrow is going out.
- Labeled connectors are represented by an identifying label inside a circle. Labeled connectors are used to bridge gaps (substituting for arrows) in the diagram for complex or multisheet diagrams. For each label, the "outflow" connector must always be unique, but there may be any number of "inflow" connectors, thus implying a junction in a control flow.
- Concurrency symbols are represented by a double transverse line with any number of entry and exit arrows. These symbols are used whenever two or more control flows must operate simultaneously. The implication of this type of connector is that all exit flows are activated concurrently when all of the entry flows have reached the concurrency symbol.

8.3.1.2 Process Mapping

Process mapping is a tool commonly used by an organization to better understand its processes. A process map is a unique type of flow charting for showing the sequence of events associated with each single event in a hierarchy of process. Simply put, process mapping is a diagram of how work flows. The mapping is intended to bring forth a clearer understanding of a process, or series of parallel processes, by showing the various activities required to produce a specific outcome and the relationship, or sequencing, of those activities to each other.

When performed at a business level, the mapping will define what the organization does, who is responsible for a given process, the standards for each process, and the criteria's for success for each outcome. The mapping will also take each specific outcome and compare its objective alongside the organization's objectives to ensure that all processes are aligned with the values and capabilities of the company (Fig. 8.1).

8.3.1.3 Deployment Flow Charts

When the process map is segmented to denote areas of responsibilities, swim lanes are used as visual references to diagram which events are assigned to that individual or group.

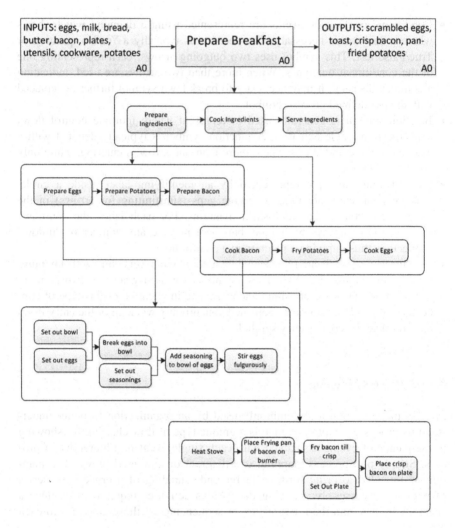

Fig. 8.1 Process mapping

8.3.1.4 Critical Pathway Charts

Critical path charts are focused on showing the sequence of events that result in a particular outcome of a service delivery system. In this version of process mapping, the specific steps required to deliver a service offering are mapped out along with pertinent guidelines that insure optimal outcomes are achieved. These essential steps are referred to as the critical pathway.

By mapping the current critical pathway for a particular service, an organization gains a better understanding of what the service offering is and how it is provided. When comparing the map as currently operated to one the shows the critical pathway leading to an optimal outcome with evidence-based guidelines, opportunities for improving the service delivery system for the offering can be seen.

Reference

1. Olson, E. M., Slater, S. F., & Hult, G. T. M. (2005). The importance of structure and process to strategic implementation. *Business Horizons, 48*(1), 47–54.

Chapter 9
Dimensions of Transformational Process Quality

Consumers normally evaluate a product or service quality during the pre-purchase, purchase, and post-purchase phases. The overall impression that they gain about an item's quality hinges on how well the firm manages the design and development of the product or service through its organizational processes, as well as how well it manages its transformational (also known as either production or manufacturing) processes. These transformational processes physically change raw inputs, such as materials, purchased components, knowledge and labor, by either changing their physical characteristics or by assembling them into finished products or services. While the consumer might be unaware of the transformational process itself, the firm should be monitoring all of the dimensions of quality for the processes that are essential to ensuring the conformity of the product or service to specifications and customer expectations.

A process is a set of interdependent tasks that repeatedly transforms inputs into a specific, measurable output. The inputs to a process typically include tools, materials, work methods, and people. Processes are not just part of the design of a work station; they are part of the inherent creative process. An activity that requires more than one task to be performed in a specific sequence so as to achieve a specific objective is a process. Processes can be seen at home, in our work, throughout the organization of a company, and even across supply chains.

All processes are subject to some level of statistical variability. Thus, output measurements will fall into a distribution pattern that can be evaluated by statistical methods. The greater the variability that is inherent in a process, the less predictable the outcomes from the process and the greater the risk of customers being dissatisfaction due to poor performance of the process output.

The design of a process is governed by two factors: the design of the product or service to be output by the process and the expectations of market demand. Because of these two factors, the dimensions of capability and capacity are of primary concern during development, management, and improvement of these processes. If adequate attention is not paid to these two important dimensions, the firm will either need to redesign the process or forgo their chosen marketplace.

© Springer-Verlag London 2015

G.N. Kenyon and K.C. Sen, *The Perception of Quality*,
DOI 10.1007/978-1-4471-6627-6_9

9.1 The Capability Dimension

The capabilities of a process are defined in terms of the process ability to consistently produce a product or service that meets the needs of the customer as defined by a given set of specifications or parameters. The assumption is that if the product meets the tolerances of the specifications then it will create customer satisfaction. A common means for articulating a process capability is to compare (through a ratio) the variance within the process to the specification limits (tolerances) of the product or service being created. The resulting ratio is called the process's "capability index." A common method for illustrating a process output is through a histogram; when the specification limits overlie the histogram, one gets a depiction of the process' capability to produce saleable items (Fig. 9.1).

Capabilities can be increased as trust-based relationships are created across the supply chain because of the increased sharing of knowledge and expertise. This increased sharing also stimulates innovation.

9.1.1 Process Capability Indices

Process capability indices (PCIs) were first introduced by the Japanese in 1974, by Joseph Juran, as a means for providing numerical measures of the ability of manufacturing process to meet given production tolerances. These PCIs indexed the output dispersion of a process with respect to the specification tolerance limits of

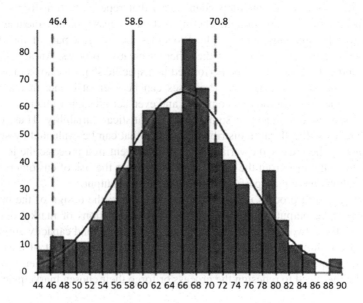

Fig. 9.1 Process capability histogram

a given process. The interpretation of the PCI gives process operators, engineers, and managers a relative understanding of the predicted percentage of output that would conform to specifications. The main advantage of these indices lies in their ability to compare process performance across product lines, industries, and countries. Another advantage is that they make changes in the process performance more noticeable, as shown in Fig. 9.2.

The two best known and used PCIs are the C_p and the C_{pk} indices. The C_p index compares process variance to design specifications in order to determine process capability. Unfortunately, this index does not guaranty that the production of materials conforms to specifications because it does not account for the location of the process with respect to specification tolerances. The C_{pk} index corrects for this shortcoming by comparing the location of the process mean to the location of both design specifications. The formulation of both of these indices can be found in Table 9.1, along with several other common metrics.

The usage of these PCIs is based upon three assumptions: (1) The process is stable, (2) the output of the process is normally distributed, and (3) the index is measuring only one performance characteristic. The most often violated of these assumptions is the normally distributed output, as many processes have non-normal or non-symmetric output distributions. Another often violated assumption is stability. Due to the issues with normality, the testing of process stability must rely on sampling and sample statistics. This can lead to additional problems. Because the C_{py} index is indifferent to these assumptions, it is recommended for most management decisions.

The interpretation of all PCIs is as follows: If the ratio is less than 1.00, the sum of the process output fails to meet design specifications; if equal to or greater than 1.00, the process is capable of production in specification materials.

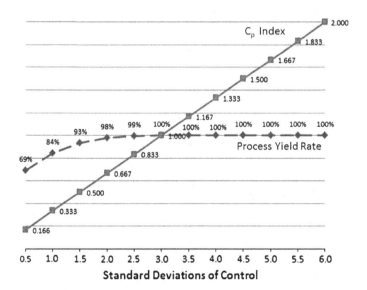

Fig. 9.2 C_p versus process yield rate

Table 9.1 Typical metrics used to assess transformational processing performance

Metric	Calculation method
Cost of goods sold	\$ Direct labor + \$ Direct materials + \$ Manufacturing overhead
First pass yield rate	$\dfrac{\text{(Number of units competed on first pass)}}{\text{(Total units started)}}$
Manufacturing cycle time	$\dfrac{\text{(Total days on hand of an item (including wip))}}{\text{(Planned shipments per day of all units containing that item)}}$
Overall equipment effectiveness	(% machine availability) × (% quality yield) × (% of optimal rate that equipment operates)
Work-in-process turnover rate	$\dfrac{\text{(Average number of units of material in production)}}{\text{(Average daily production demand)}}$
Defects per million opportunities (DPMO)	(Average number of defects/unit) × (number of opportunities) × $(1,000,000)$
Process capability (C_p)	$\dfrac{\text{(Upper specification limit)} - \text{(lower specification limit)}}{\text{(6 standard deviations of process variance)}}$
Process capability (C_{pk})	$\text{Min} \left\{ \begin{array}{l} \dfrac{\text{(process mean)} - \text{(lower specification limit)}}{\text{(3 standard deviations of process variance)}} \\[2mm] \dfrac{\text{(upper specification limit)} - \text{(process mean)}}{\text{(3 standard deviations of process variance)}} \end{array} \right.$
Process capability (C_{py})	$\dfrac{\Phi^{-1}(\text{Yield})}{3}$

9.2 The Capacity Dimension

Customer satisfaction is also linked to the company's ability to deliver products and services in the quantities requested at the time and place contracted for. Thus, the capacity of the firm's production processes is critical to meeting those demands. Capacity is a dimension that is predominantly focused on volume per given time period. It can be defined as the maximum volume of work that can be completed within a specific time frame. Capacity management on the other hand seeks to ensure that the capacity available to the firm at any given time can meet the capacity required for demanded goods and/or services. The key determinants in capacity planning are market demand, product design, process design, and process capabilities.

The management of capacity is an operational issue as well as a strategic, organizational issue and often involves costly trade-offs. As such, it requires both long-term planning and short-term control. Long-term objectives with respect to anticipated market demands and the company's ability to meet those demands need to be established as a first step. The guidelines set by these objectives form the basis for when and how much capacity the firm will need, as well as the operational strategies for what type of capacity will be used. These include factors such as internal production, joint ventures, and outsourcing. The importance associated with this level of planning is rooted in the fact that even if the firm has the best product or service available and if delivery commitments to customers are not met, the company will lose its competitive advantage and customers will defect to the competition.

There are several issues that affect the capacity decision: variability, capacity utilization, and inventory. Variability affects process capacity in two ways: manufacturing cycle time and conformance to specifications. Capacity utilization is affected by material flow rates, market demands variability, and scheduling.

9.2.1 Supply Issues in Capacity Management

Competitive advantage is a fleeting commodity and it must constantly be reestablished. Not only must the firm ensure that its internal processes are capable of adding the value sought after by the customer, it must also guarantee that the materials, components, and support services that are the inputs to their processes are available and meet or exceed specifications. So, as management struggles to achieve and maintain the company's competitive positions, they must also consider the design of the firm's supply chain. Disintegrating and reintegrating the organization's supply chain structure are often necessary to ensure that the requisite input needs are met.

Recently, my wife and I purchased two new chairs for our living room (pictured in Fig. 9.3). We were looking for chairs that incorporated the lounge room

Fig. 9.3 Chairs

comfort, while presenting classic old fashion style and character. The store that we purchased the chairs from told us that there was a 4- to 6-week delivery window. This is typical when ordering from the manufacturer. Sixteen weeks later and several increasingly frustrated conversations later, we finally got the chairs. The apparent reason behind the late delivery was that the manufacturer (located in Tennessee) was employing undocumented, foreign workers, and the US Immigration Service raided the factory. Even though I do not blame the workers for wanting good paying jobs and even understand the competitive cost pressures that drive the manufacturer to hire undocumented labor, as a customer I was not happy with the unexpected and unreasonable long delay. The manufacturer not only failed to manage the risks to his production lines, he was operating illegally. The retailer, although providing quality products at competitive prices, failed to ensure the integrity of his supply chain. Even after considering the myriad issues involved, as a customer I was not happy. I was expecting to receive the products I purchased within the agreed upon terms of the contract which were not met by the retailer/manufacturer. In all probability, I will not purchase from that retailer or the manufacturer in the future.

The success of any firm in the current business environment hinges upon its ability to coordinate and synchronize the network of business relationships that constitute its supply chain. The principle objective is to enhance operational efficiency, profitability, and competitive position of both the firm and its enterprise partners. The problem is that supply is often constrained, a situation that has become even more difficult with the globalization of marketplaces. In addition, it is more challenging to assess market conditions and ascertain how to allocate scarce supplies so as to achieve your goals. Factors that contribute to the problem include the following:

- The large number of nodes in the supply chain.
- The increasing number of silos of information and the increasing complexity of the information needed for accurate analysis and modeling.

- Too many disparate systems involved with the planning, execution, and monitoring of supply chains.
- The multiple levels of planning (strategic, aggregate, and operational) and the breadth of planning needed (domestic, multinational, global).

Consider the following scenario: Japan recently experienced a devastating earthquake and tsunami. Your company is looking to move into Asian markets, and you have been asked to assess the robustness of possible supply chains for dealing with disruptive risks. What previously looked like sound strategies could create major macro-level vulnerabilities when subjected to large-scale disruptions. Practice has also shown that trying to hedge your risks by sourcing every component from multiple locations is very costly. In these cases, what is the correct option?

With the ever-increasing rate of technical change, abrupt economic shifts, and new tactics of competitors, individual capabilities can be lost overnight. To successfully deal with all of these issues, there are five qualitative factors that must be addressed when evaluating capacity and capability strategies. These are customer preferences, the rate of change in technology, competitive position, supply base capabilities, and system architecture.

9.2.2 Inventory Issues in Capacity Management

Inventory management involves the rationalization and control of raw materials, components, semi-finished parts, components, and assemblies, and finished goods waiting to be sold, which are transported or used at a point of the supply chain. The goals of inventory management are as follows:

- Improving customer service levels.
- Reducing overall logistics costs.
- Buffering processes against random fluctuations, or disruptions, in demand and lead times.
- Managing availability of seasonal items.
- Speculating on price patterns, etc.

When demand is high, excess production is easily consumed. However, when demand drops-off, inventories can build up rapidly. The longer the supply chain, the slower the reaction times to shut down production and the greater the financial impact.

The root causes of most inventory issues are as follows: demand variability, extended lead times, the lack of visibility along the supply chain, lack of collaboration between supply chain participants, and the quality and reliability of suppliers. These factors are also major contributors of variability in the supply chain.

In the past, organizations and supply chains decoupled their various functional and geographic components through the usage of inventory buffers. In today's fast-paced business environment where "time to market" and flexibility are major factors in long-term competitiveness, ignoring the functional and organizational interdependencies of the supply chain is no longer feasible. However, the continuing uncertainty of demand, and/or supply, and/or production cycle times, makes it necessary to still hold inventories.

Traditionally, in make-to-stock production systems, holding safety stocks was the remedy for demand variability. Where demand was discontinuous, as in seasonal markets, adjusting production capacity is an often used solution. Risk pooling and postponement strategies have been used to help resolve the trade-offs associated with inventory buffers in the supply chain.

Research has found that integrated capacity and inventory management approaches can outperform decoupled tactics. The availability of flexible capacity offers great potential when contingent capacity is relatively low cost, or when backordering costs are high, or when the fixed costs of production are high, or when existing capacity is limited. As volatility of demand increases, there is more value in flexibility and greater support for establishing long-term contracts with contingent capacity providers.

9.2.3 Reliability Issues in Capacity Management

Process reliability is defined as the probability of defect-free performance over a given time frame under specific operating conditions. Usually, a process is most reliable when operated at a constant speed, under a constant load, and under a set of constant conditions. A process is at its "Optimal Operating Condition" when it is performing under the previously defined circumstances, and those conditions result in its lowest operating costs. These operating costs include material costs, machine costs, labor costs, maintenance and repairs costs, utilities, space rents, etc., allocated at the cost per unit.

Reliability, like quality, cannot be added after the fact; it must be designed into the process. Even with the best design, if the process assets are not installed correctly (e.g., proper foundation, leveling and alignment), they will not perform as planned. In addition to proper installation, operators must be thoroughly trained in the proper operation of the machines that they are responsible for (e.g., setup, startup, ramp rates shutdown, and maintenance). Furthermore, if the equipment is not maintained properly, its reliability along with its capabilities will diminish over time.

Process reliability requires that equipment is effectively utilized. All physical assets are designed to perform specific tasks within a given range of conditions, called the "operating envelope." This envelope defines the limitations on continuous operations, the number and frequency of starts and stops, operating speeds, temperatures, and other applicable setting and conditions. In addition, the range of inputs with respect to outputs is an important factor. All these variables have implications for the scheduling of work on the process, product selection, batch sizes, changeovers, processing time, maintenance, etc.

One of the main principles of total quality management is continuous improvement. The biggest enemy in manufacturing is variability. Thus, ongoing efforts not only strive to improve process' throughput rates, but also to reduce variability of outcomes. In order to maintain the reliability of a process, it is essential that a rigid change management process be in place to ensure the final fit, form, and function of the process. It has been suggested that up to a quarter of all reliability problems are the result of uncontrolled modifications and changes to the processes.

9.3 The Flexibility Dimension

Flexibility is another important dimension in the design and management of production processes. Garvin [1], a prominent operations management professor and researcher, has defined process flexibility as "the ability to respond effectively to changing circumstances" and that "one manufacturing process is more flexible than another... if it can handle a wider range of possibilities." From the perspective of manufacturing managers, flexibility is a function of the process range and response. Here, range flexibility is the total range of capabilities that the production system is capable of achieving, and response flexibility encapsulates the different ways that changes (e.g., setups) can be implemented in the system.

Because of the dynamics of the marketplace (e.g., changing customer expectations, competition), the firm is always concerned about whether process designs are flexible enough to meet the changes of future market demands and requirements. If they are not, then the firm will again be faced with either a total reengineering of its processes or the forgoing of market opportunities. If processes are flexible, then less expensive modifications and improvements can be made as needed. There are several types of uncertainty that drive the need for flexibility: Shifts in market demand patterns drive the need for product mix flexibility; the uncertainty about product life cycle durations drives the need for changeover flexibility; the uncertainty about equipment failures and repair durations as well as the uncertainty about input arrival rates drives the need for alternative process routing flexibility; the uncertainty of aggregate demand drives the need for volume flexibility; and the uncertainty about incoming material qualities drives the need for material flexibility. Instead of staging product lines for each product offering, flexible process designs offer a relatively economical means for addressing these uncertainties.

There are five basic levels of flexibility: the individual asset (e.g., worker or machine), the type of manufacturing (e.g., forming or assembly), the process, the factory, and the supply chain. Supply chain flexibility covers several of the uncertainties that operations manager routinely face. An effective supply chain can reduce the uncertainty of incoming material quality, thus reducing material flexibility needs. An effective supply chain can also reduce the uncertainty about on-time deliveries, thus reducing the need for sequencing flexibility. It is also possible that supply chain flexibility can reduce the need for changeover flexibility.

With new process management techniques, such as lean manufacturing, and increased equipment capabilities and improved supply chain effectiveness, lead times can be reduced making forecasting more accurate. This reduces the need for mix and volume flexibility. With the advent of preventative maintenance programs, the need for alternative routing flexibility is reduced. Over the past two decades, there have been significant changes in how factories have aligned themselves (e.g., Original Equipment Manufacturers, outsourcings, etc.) and how they compete, further reducing many of the uncertainties that drove the need for flexibility. These same changes have also increased other forms of flexibility, such as modification flexibility (e.g., product variety), volume flexibility (e.g., production location), and changeover flexibility (e.g., new product development and commercialization).

A facilitating element to the dimensions of flexibility, capability, and capacity is technology. As product functionality and capabilities increase, the difficulty of manufacturing the produce products increases significantly. Advances in manufacturing technology have enabled production processes to produce leading edge products in cost-efficient ways. This in turn has led to business success and conversely the wrong technology. Even if the appropriate technology is selected, but implemented incorrectly, a loss of competitive advantage is likely to occur.

Another facilitating element involves product design. As the competitive pressures from globalization have increased, there has been an increased emphasis on the new product development process to design easy to manufacture products of great variety. As a result, both product design and process development have promoted greater manufacturing flexibility. The changes in product architecture and modularity have facilitated the development of assembly lines that are capable of producing multiple products. They have also led to the establishment of work cells that can match the pace and mix of such assembly lines. The variety of products produced by these changes has promoted greater mix flexibility, changeover flexibility, and modification flexibility. At the same time, easier manufacturing processes have led to greater rerouting flexibility, volume flexibility, material flexibility, and sequencing flexibility.

There are four distinct methods to increasing process flexibility: design, deviation, under specification, and change. The design methodology defines both a primary execution path and alternative execution paths for accomplishing the objectives of the process. The deviation method allows for deviations to the defined execution paths. These deviations may encompass changes at a specific process instance, but are not allowed to change the process model. The under-specification method involves an execution path that is only partially defined and is completely defined at the time of execution based upon the needs of the item being processed. This type of flexibility model is most suited to processes where adjustments that will need to be made at specific points in the process are known ahead of. The change method to flexibility allows for the modification of the current execution path to a new path based upon the process instance. This often happens when unforeseen events occur during processing.

9.3.1 Flexibility Issues in Supply Chains

There are two basic types of supply chains: efficient process chains and flexible process chains. For markets that emphasis large volumes of relatively stand products at low cost, there is little need for process flexibility. As markets increase their emphasis for more specialized products, developing processes with greater flexibility that can maintain low production costs and provide high quality becomes more imperative.

The degree of effectiveness of the supply chain depends upon the integration of several business processes across the chain (see Fig. 9.4). The greater the degree of integration between customers, suppliers, and the firm, the more effectively the supply chain can compete in highly competitive markets. One of the first requirements in designing a flexible supply chain is to understand the type of product you are offering and the needs of your customers. Next, you must align the appropriate resources (and supply chain partners) in a structure that can meet the customer's needs. With the right incentives and a high level of trust between partners, the supply chain will be able to adapt to shifts in the market and have the ability to respond to short-term changes in demand or supply smoothly.

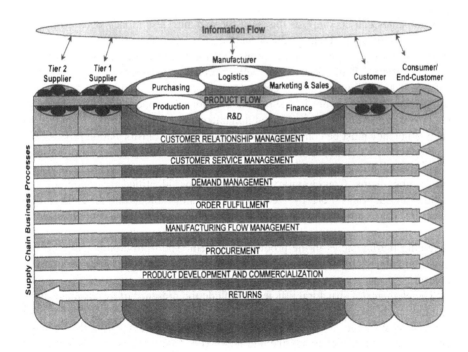

Fig. 9.4 The business process for supply chain integration (Published with kind permission of © Emerald Group Publishing, per the STM permissions guidelines. Cooper et al. [2]. All rights reserved)

In designing and executing a supply chain strategy, there are three key aspects to creating flexibility: the buyer–supplier relationship, demand, and manufacturing. The buyer–supplier relationship aspect is focused on the procurement of input materials and services to the firm's manufacturing systems. The demand aspect is fundamentally about creating value for the customer. The manufacturing aspect of supply chain management is concerned with strategic flexibility and speed to market of products flowing through the supply chain as a production system.

9.4 The Efficiency Dimension

The importance of an efficient production and delivery system is rooted in economics as noted by its definition. Efficiency is measured as the ratio of actual output (e.g., units produced) to actual inputs (e.g., labor costs, material costs, equipment costs, utilities, rent). All companies have a few things in common regardless of industry; they operate similar equipment, utilize similar resources, and face similar constraints as their direct competitors. Furthermore, at the end of the day, they still need to have produced a profit in order to maintain and grow the business.

Companies can accomplish this goal by spending less money during the creating and delivery of their products and services than the price with which the market is willing to pay for those items; thus the need for efficiency in their production systems. If the firm cannot deliver products and services that customers value and are willing to buy, it cannot remain competitive and thus will go out of business. Quality is not just about the creation of a superior performing product, and it is also about the creation and maintenance of competitive advantage. As such, efficiency must be considered a key dimension of transformation processes.

Production efficiency is defined as occurring when the production of an item is achieved at the lowest input cost possible, with respect to the production levels of other items. Contemporary researchers have identified three components within production efficiency: technical efficiency, allocation efficiency, and structural efficiency. The first two components are generally accepted as being within the firm's control. Thus, improvements in efficiency can be made by removing inefficiencies in the production system. The third component involves environmental elements that are external to the firm and thus outside of their control.

The elements of structural efficiency include the presence of government intervention, such as regulations, rationing, and dictated corporate restructuring like what General Motors Corporation experienced in the bankruptcy after the financial crisis of 2008. These elements include factor (i.e., input materials and resources) adjustment costs. The inability of the firm to adjust to changes in these elements over the short term reduces its efficiency of factor mobility.

Technical efficiency is concerned with the speed and effectiveness with which a given set of inputs is used to produce an output. As an example, a manufacturer is considered to be technically inefficient if it employed more workers than was absolutely necessary to produce a given set of outputs. Allocation efficiency, on the other hand, is associated with the distribution of those goods and services that are the most desirable to society and also are in high demand. In other words, the point at which the marginal benefit is equal to marginal cost. For example, perfectly, competitive markets are considered to be allocatively efficient, whereas monopolies, monophonies, externalities, and public goods are not. Thus, managers attempt to optimize the allocation of their resources within the constraints of their external environment.

9.5 The Conformance Dimension

Ultimately, the product or service being produced by the firm's transformation processes must routinely execute the functions required or expected of the product as designed by engineering and demanded by the customer. Even though product design addresses every functional concern of the customer, if the product is not produced correctly, it will not perform as expected. Even if all the components and assemblies of the product conform to specification and initially the product performs as expected, if the materials and/or workmanship was defective in any way, the product will not continue to perform at expected levels throughout its service life. Premature failures, or excessive maintenance, will result in customer dissatisfaction and the loss of competitive advantage.

Taguchi explained this loss of consumer satisfaction due to variance from nominal specifications with his quality loss function theory. He believed that across all industries, in order to extract the maximum performance from a product, it needs to produce an outcome on target (i.e., the nominal specification). The greater the variance from that nominal specification, the poorer the manufactured quality and that reacting to individual items inside and outside specification was counterproductive.

Taguchi argued that the engineering community had to understand the quality costs associated with various situations. He also insisted that engineers needed to widen their horizons beyond just the consumer to considering the *costs to society* for these poor-quality results. To better understand the cost associated with variance from specifications, he explained that even though the short-term costs may simply be those scrap and rework, the long-term costs of any manufactured item that do not meet the nominal speciations would also result in losses to the customer and to society because of early wear out or failures, difficulties in interfacing with other parts, or the need to build in safety margins.

Because of the difficulty in collecting the post-purchase costs associated with poor quality, manufacturers usually ignored them in their product development decisions. Taguchi believed that this practice was socially irresponsible because the post-purchase costs prevented markets from operating efficiently. He also believed that post-purchase losses would inevitably find their way back to the

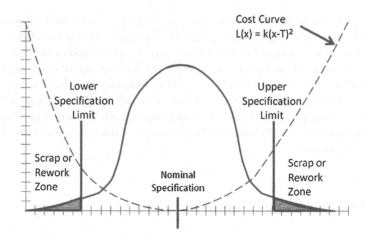

Fig. 9.5 Taguchi's quality functional loss curve

originating company. By working to minimize them, the firm could enhance their brand reputation, win markets, and generate greater profits. Taguchi theorized that the costs associated with poor quality were proportional to the square of the distance (i.e., variance) from the target value times a constant cost factor. This concept is illustrated in Fig. 9.5.

9.6 The Inventory Management Dimension

Inventory is the two-edged sword that makes the system work. There are three principle objectives associated with inventory management: (1) to minimize the risk of system failure due to unplanned interruptions, (2) to control the firm's investment in inventory, and (3) to minimize the total cost associated with inventory. Inventory management helps the manufacturing process by placing buffers at strategic points throughout the supply chain. Thus, it ensures a constant supply of materials and other inputs to the production system during an interruption in supply and equipment failure and also helps meet minimum customer service levels. The downside on maintaining these inventory buffers throughout the supply chain is that they not only add expenses that reduce profit margins, but also encumber the money that the firm needs for investments in their products and business processes.

There are several aspects to the effective management of inventories: How much inventory and which particular items should be held; where will those items be held and at critical times should these inventories be replenished? To complicate these decisions, there are several stakeholders all of which have different agendas and objectives. Buyers want to purchase inventories at the lowest cost. This could result in risks to suppliers that affect their ability to consistently provide items that meet specifications without interruptions. Manufacturing wants inputs that consistently meet specification that are delivered in just the right amounts to the right locations. Sales want as many items in stock and as much variety as possible all of the time. Supply chain planners want to maintain the balance between supply and demand. Finance sees inventory as an expensive but necessary evil. Executives see inventory as a means of meeting strategic objectives. Customers want what they want, when they want it, and they want it at the lowest price. Given these varied demands, inventory managers are hard put to keep everyone happy.

Despite the improvements in technology and management techniques, supply glitches occur for any number of reasons. These glitches cause demand variability throughout the supply chain and have been reported as to causing as much as a ten percentage reduction in shareholder value. The more removed from the initial glitch point, the greater the demand variable. This variability can cause significant problems in the material flows across a supply chain. This in turn will drive inventory policies into more conservative risk adverse positions and will also drive costs up. The ultimate solution to these demand-driven problems involves both lead-time reductions and greater visibility of true customer demand patterns across the supply chain. Other management actions could also involve higher safety stocks and reduced responsiveness to changes in customer demands.

Inventory management is a process that involves recording and monitoring of stock levels, forecasting and demand management planning geared toward determining what, when, and how to ordering of replenishment stocks, as well as the identification of new stock items. Inventory management is a vital part of any business because of the large expenses associated with stocking inventories. In addition, an ineffective inventory management system can have a negative impact on customer satisfaction.

Inventory availability has a direct impact on the firm's competitiveness and its profitability. Often, customers will be more tolerant of out-of-stock situations when their purchases are of a discretionary nature or when there are no acceptable substitutes available. Even then, when products are consistently unavailable, there can be a serious risk of brand defection.

There is an opportunity cost associated with customers opting for alternative product solutions. The costs to the customer will include the time and expenses of searching for alternatives to their unavailable choice, as well as the potential loss of quality and/or functionality. The costs to the firm include not just the lost sale, but the potential for lost customers in the future. Multinational and global organizations can be seriously affected by lost future business due to customer behaviors differing from one country to another. For example, European customers are more likely to switch brands than favored stores, while American customers are more willing to accept alternate package sizes or a different product within the same brand family or to even switch stores altogether.

There are several functional elements that affect inventory control systems: customer order fulfillment time, returns, pricing, and accurate inventory records. The objective with customer order fulfillment times is to keep fill times as low as possible. To accomplish this objective, a thorough understanding of demand patterns for the various offerings is needed. Popular items may be stocked in accordance with a service-level policy, while a limited amount of special order items might be kept. The objective with a returns policy is to maintain a high level of customer satisfaction. In retail environments, having standing agreements with manufacturers for the swap-out of damaged or malfunctioning items is needed. For the manufacturer, maintaining extra inventory equal to the expected product failure rates during warranty periods is recommended. Carrying excess inventory or carrying inventories for protracted periods of time can have a significant and detrimental effect on operating costs and marketing pricing. Maintaining accurate inventory records along with accurate data on market demand patterns is essential for meeting the objectives of these functional elements.

9.7 The Cost Dimension

Warren Buffet[1] in his letter to shareholders stated that the intrinsic value of a business is equal to the discounted value of the cash flow of the business over its remaining life. He further stated that the challenge of managers was to implement strategies than can increase the free cash flow of the business either directly or indirectly. Thus, independent of all of the other dimensions discussed, if the firm's production processes cannot produce and deliver products and service cost effectively, the company is out of business. Costs are a major factor in the customer's perception of value. Furthermore, when expenses are lower than that of competitors, a greater opportunity exists to achieve higher sales and still remains profitable. A company can therefore be a market leader by being innovative in cost control. As discussed previously in the theories of quality, an effective quality management program can have a significant positive affect on both market pricing and operating costs as shown in Fig. 4.1.

Establishing and maintain an effective supplier management program can also help lower operating costs. Suppliers possess extensive knowledge about the potential of the products and services they provide. Thus, they can add significant value to their customers in various ways. Activities such as just-in-time can help reduce inventories. This in turn will reduce the amount of working capital, and the firm needs to reserve for operations and improve cash flows.

9.8 The Order-Handling Dimension

The importance of order handling is related to time and personal involvement. Order handling is not just a sales process; it is also a business planning and an inventory process. If a company cannot produce a product within the customer's lead-time expectations, then it must forecast their expected sales and produce the forecasted items ahead of demand and stock them. This type of uncertainty also drives costs in

---

[1] W.E. Buffett, Chairman of the Board, Letter to the Shareholders of Berkshire Hathaway Inc. Dated March 7, 1995.

the inventory management areas because of the need to carry safety stocks of each product and the additional carrying costs or scrap costs associated with over production or the loss of customer goodwill associated with under production.

9.9 The Input Dimension

The importance of input quality is mostly a customer issue as opposed to a consumer issue. If the supplier maintains a high level of quality, then the customer does not need to verify quality before the purchase or deal with returns or warranties after the purchase. But if the supplier's quality is suspect, additional costs are going to be incurred due to the need to monitor incoming purchased items so that they meet specifications before they are accepted and placed into inventory.

9.10 Implications to the Customer

From an operational perspective, variability is the enemy. Variability drives uncertainty, and the uncertainty of outcomes will drive managers to implement safeguards to ensure their ability to meet market demands. Though consumers and

Dimensions of Transformational Process Quality :	Customers			
	Investors	Employees	Consumers	Society
Conformance		✓		
Capabilities	?	✓		
Flexibility		✓		
Efficiency	?	✓		✓
Inventory Mgmt.		✓	✓	
Order Handling		✓	✓	
Cost	?	✓	✓	✓
Incoming Inspection		✓	?	

Fig. 9.6 Visibility of process quality dimensions to the customer

most other customers will never see the firm's actual production processes, they
do see the results of those processes. They also see how the firm's delivery systems perform. Variability in the performance of products or services will dilute
customer satisfaction and reduce the firm's ability to maintain its competitive
advantage. The safeguards that management put in place to hedge the effects of
variability cost money and can in some cases reduce operational capabilities. All
of these actions negatively affect pricing and have a deleterious effect on customer
satisfaction (Fig. 9.6).

References

1. Garvin, D. (1987). An agenda for research on the flexibility of manufacturing process.
   International Journal of Operations and Production Management, 7(1), 38–49.
2. Cooper, M. C., Lambert, D. M., & Pagh, J. D. (1997). Supply chain management: More than a
   new name for logistics. *International Journal of Logistics Management, 8*(1), 2.

Chapter 10
Process Improvement Methods and Tools

The productivity of its transformational processes (i.e., production systems) has an enormous impact on the firm's ability to deliver quality products and services to consumers. External customers rarely see a company's production systems and in most cases are not interested in its details. However, being able to get the products and services they desire in a timely fashion greatly affects consumer perceptions of those products and services. Even if the product or service is perceived as being of good quality, the inconvenience of not getting it in a timely fashion reduces the customer's since of value and often causes dissatisfaction. With respect to internal customers, the higher the effectiveness and efficiency of the production systems, the greater their sense of value and perception of quality.

One of the harsh realities of life is that the strong survive and the weak die. We live in a highly competitive world, and just because you are on top today does not mean that the same situation will exist tomorrow. The degree of competitiveness in virtually every market has significantly increased with the invention of the Internet and the associated improvements in telecommunication capabilities. As a result, survival in business means embracing change and leveraging it toward competitive advantage.

The primary mechanism for a company to stay at the forefront of the customer's thoughts is through its products and services. Given this goal, management needs to design processes that are capable of producing product and/or delivering services that meet customer expectations. The more effectively and efficiently those processes transform inputs into outputs, the higher the company's productivity, and the more competitive it will be in the marketplace.

The two basic tenets of quality management are as follows: continuous improvement and customer focus. Assuming that marketing has correctly identified and translated customer's requirements for products or services into specifications and that engineering has accurately and innovatively translated those specifications into designs, manufacturing needs to design transformational processes that can efficiently and effectively produce those designs into products and/or services that will

© Springer-Verlag London 2015
G.N. Kenyon and K.C. Sen, *The Perception of Quality*,
DOI 10.1007/978-1-4471-6627-6_10

satisfy the customer's needs and expectations. Furthermore, assuming that all of this occurs in a timely fashion, there are no guaranties that the outputs of these processes can maintain the company's competitive advantage overtime.

10.1 Continuous Improvement

In order to remain competitive, companies must continuously improve their processes. In the manufacturing and service delivery environments, these improvements involve the reduction of variance and increasing both efficiency and effectiveness. These types of improvements often result in increased capacity, productivity, flexibility, and responsiveness; thus, leading to greater profitability.

10.1.1 Reducing Variability

Variance is defined as any amount of deviation from the nominal specification. There are numerous problems associated with variance. When there is variance from the nominal specification in a part's production, there is a corresponding loss in product performance and/or scrap or rework of the component. Variance increases the unpredictability in the product's performance, in quality, and in manufacturing throughput. Variance in a process will reduce its capacity. Process variance also reduces the system's ability to detect problems and increases the difficulty of discovering the root cause of the problem.

There are four major sources of variance in production:

1. Insufficient design tolerances due to poor design practices and/or unrealistic or incorrect requirements.
2. Inherent variability of manufacturing processes (i.e., manpower, methods, materials, and machinery).
3. Inherent variability in measurement systems.
4. Variability in purchased components.

Two ways of dealing with process variance are as follows: (1) through continuous and systemic improvements to the process, and/or (2) improved robustness of product and service designs. Studies have shown that about 75 % of manufacturing problems come from the design function, and only about 80 % of those issues are detected in-house. The design of the final product also influences the design of the production processes needed to create the product or deliver the service. Thus, taking a holistic view of the design process in conjunction with performing a value analysis of product designs and manufacturing process designs is one of the first steps in understanding the potential source for variation and the consequent steps to reduce it.

10.2 Improving Production Effectiveness

Effectiveness is determined by comparing what a process is theoretically capable (i.e., design capacity) of producing at a given level of inputs, with what it actually produced. With respect to production systems, effective capacity is determined by subtracting allowances such as personal time, maintenance, and scrap from the system's design capacity. When seeking to improve a system's capacity in the short term, management must work with a system's strengths and avoid its weaknesses. To be effective in the long term, you must change the system's weaknesses into strengths.

10.2.1 Measurement Problems

The first hurdle to overcome when trying to improve the effectiveness of manufacturing is to measure it correctly. Frequently, companies will measure effectiveness using a "standard" cost system. In this type of system a standard is set, defining both the time and cost of each item produced by the company. A variance is then determined by comparing actual times and costs against the standard time and cost of the actual number of units produced. If actuals are less than the standard, then the variance is favorable. If actuals are greater than the standard, then the variance is unfavorable. This sounds good in theory. However, it has to be understood that the perceived effects of this localized behavior can be caused by the measurement system.

Frequently, the driving forces in the development of measurements and standards are assumptions about the behavior of the system. The typical assumptions driving most companies' usage of production metrics are as follows:

- The company has excess capacity available.
- The standards used by the production system are accurate, i.e.,

  - Processing times are accurate.
  - Setup and batch sizing are correctly taken into account in the standards.
  - Labor, material, and overhead costs are accurately reflected in the standards.

In reality, for most companies and most production processes, the following is often true:

- Excess capacity may or may not be available, due to market demand fluctuations.
- Processes and work methods are constantly being improved (i.e., take less time).
- Labor rates will vary depend upon the pay rate of the personnel used to perform the tasks.
- Material costs will be dependent on inflation and changes to material specifications.

- Overhead charges may have changed since the standard was set due to organizational changes (i.e., size, structure, and alignment).
- Time standards for determining performance are often based on averages or estimates and are often inaccurate. This inaccuracy is exacerbated further by time, changes in processes, and lack of maintenance and verification.

The issue is not that there are standards, but that managers are driven overtime to be increasingly "efficient." Furthermore, efficiency tends to be localized in nature and often causes collateral effects across the business. In order to better understand these effects, take the following example.

The ACME manufacturing company produces custom widgets. The design capacity for their production process is 2 custom widgets per hour. Their effective capacity is rated at 1 part per hour. The standard cost of labor is $20 per man hour, and the standard cost of machine time is $50 per machine hour, and the standard material costs for a custom widget are $5 per unit. In this system, if AMCE produced 100 custom widgets for the measures found in Table 10.1.

If production has unfavorable performance against standards, it must produce widgets in excess of demand in order to earn back the hours lost on the parts that had underestimated standards. They do this by focusing on those parts with overestimated standards. As a result, production managers will tend to avoid producing parts with underestimated standards in the future, unless it is necessary to satisfy customer demand. The overall impact is that production managers will perceive that they are being punished by producing to demand, and being rewarded when producing parts that are not in demand for inventory. The implications of this scenario is that production variance reports will look good, but the company will have cash tied up in inventory that is not essential to meeting customer demands.

One mechanism for minimizing the negative effects of producing underestimated parts is to postpone production on them until there is a sufficiently large batch. This will reduce the number of setups typically needed for supporting customer demands, but also the cost variance. This type of solution results in slower response to customer demands, thus creating customer dissatisfaction. In addition, this solution will drive up inventory levels and inventory carrying costs. Another

Table 10.1 ACME manufacturing company production standards and costs

Parts produced	Standard machine hours	Hours earned	Hours worked	Variance
100	1	100	110	(10)
	Standard labor rate	Labor earned	Labor spent	Variance
	$20	$2,000	$2,200	($200)
	Standard material costs	Materials earned	Materials spent	Variance
	$5	$500	$550	($50)
	Standard overhead costs	Overhead earned	Overhead spent	Variance
	$1	$100	$110	($10)

disadvantage here is that when lead times get too large, managers will feel the need to order overtime in order to reduce customer dissatisfaction at the expense of additional cost variances.

Another consequence of this scenario is the negative impact on labor flexibility. In plants with multi-tiered pay structures, managers will tend to not use higher paid, skilled workers on jobs where the labor standard is underestimated; thus, limiting the type of workers that can be used on those tasks. The delays associated with this behavior can result in late deliveries to customers. It can also reduce interdepartmental cooperation due to the negative cost variances ascribed to using higher pay workers on low margin tasks. A third behavioral effect is that managers need to keep everyone working in order to keep labor utilization high. Thus, high-paid workers will often be working on tasks that are not needed (i.e., producing parts for inventory); thus, driving up inventory levels and inventory carrying costs.

The only way to improve system performance is to eliminate those items and/or activities that produce bad behaviors. One suggested method is anytime there is a change to a process, the standards associated with measure productivity in that process must be recalculated in order to account for the effects of the change, and verified to be accurate.

10.3 Improving Production Efficiency

Efficiency is a quality-based metric. It measures how well the system converts inputs into outputs. To determine efficiency, one must compare the level of actual output of a system to the expected output, given the amount of inputs. In most cases, the level of expected output is the system's effective capacity.

Production is a flow-oriented activity. As such, a few common steps can be used to improve its efficiency. The first step is to maximize material utilization. This step involves studying the flow of materials through the production system, eliminating the waste (i.e., motion and idle time) in those flows. Here, one has to identify the bottlenecks and institute appropriate controls to manage the flow through these areas. In addition, the use of automation to execute routine and mundane activities will free up labor to deal with activities that are more complex.

The next step is to improve the overall equipment effectiveness (OEE). This metric is computed by the product of three operational factors: equipment availability, performance, and quality. OEE should be benchmarked, improved, and optimized.

The third step is to implement continuous improvement programs. High-yield areas for improvement are as follows: preventative maintenance, automation, lean manufacturing principles, supplier audits, benchmarking, and training. Finally, the firm has to also control costs and simultaneously find ways to reduce them while continuously improving all aspects of the process.

10.4 Improving Productivity

In many cases, it may not be cost effective or strategically necessary to improve the production system. However, the system's labor productivity might need to be improved. As a measurement, productivity represents the system's average efficiency. It is normally calculated as the ratio of the system's output divided by its demand. The importance of this metric for management is that it directly affects the company's profits.

In managing labor productivity, there are a few things that management should strive to avoid and several things they should strive to do. The "No-No's" in managing productivity revolve around behavioral issues that tend to hinder employee productivity: micromanagement, long meetings, and irrational or dysfunctional thinking.

The Merriam-Webster Dictionary defines micromanagement as "to manage especially with excessive control or attention to details." Managers that engage in this style of management justify it as being attentive of detail and being hands-on. In fact, they are either obsessive control freaks or they feel that they must continuously drive others to success. The problem is that it they are conveying a level of distrust to the employee and as a result will lower his confidence. They are in affect disempowering the employee. The warning signs that someone is becoming a micromanager are as follows:

• They resist delegating work.
• They tend to immerse themselves into overseeing every aspect of a project or in the overseeing of projects that others are engaged in.
• They become detail hounds; constantly correcting every tiny error.
• They take back work that is delegated out if mistakes are found.
• They start discouraging others from making decisions without prior consultation.

From an organizational perspective, micromanagers are constantly reaffirming their approach every time they find a mistake; thus, intimidation employees and reducing their confidence levels in their own abilities. If the employee is a strong, confident individual, micromanaging them will destroy their motivation overtime. The result is that the organization is being setup for failure.

Good managers will seek to empower their employees, as well as those around them, by giving them opportunities to excel and succeed. There are both structural and behavioral things that management can engage into promote higher levels of labor productivity:

• They can clearly define the company's strategic goals and objectives, and then link the various activities within the company to those goals and objectives. These goals and objectives need to be achievable, manageable, and measurable.
• They can increase employee engagement by striving to educate and inform employees on how they are currently contributing to the company's success and how they can increase that level of contribution in the future.

- Rotate employees through various jobs and assignments during their tenure with the company.
- Develop the organizational culture to promote and support innovation.
- Ensure that employees are properly trained and support for the tasks they are assigned, including being provided with the proper resources for the task. Furthermore, managers need to routinely follow-up with employees to insure the work in progressing on schedule and to head-off problems.
- People need to be held accountable for their actions and their decisions.
- Keep the lines of communications open between employees and their managers. Bosses need to appear humane (i.e., open, caring, and solicitous of feedback and new ideas).
- Continuously motivate employees.
- Review the company's recognition and reward policies to insure that they are aligned with current goals and objectives.

10.5 Balancing Effectiveness, Efficiency, and Productivity

There is an upper limit to the amount of efficiency gains that can be achieved with any process before the process has to be totally redesigned. The law of diminishing returns also comes into play with improving process efficiency. Frequently, efficiency can be improved by doing thing right the first time. Efficiency can be improved by improving the control of the process, so that doing things right is systemically driven (i.e., mistake proofed). It is also improved by reductions in process variability.

Similarly, effectiveness has an upper limit. That limit is controlled by the design of the process. Often effectiveness can be improved by increasing the speed of the process. The question then is, "What is the optimal speed for the process?" The concern with increasing the process speed is related to increased costs of operations and decreased quality. When does increased process speed decrease efficiency? By balancing effectiveness with efficiency, productivity can be increased, thus increasing the internal customer's satisfaction.

Consider the following production system that shown in Fig. 10.1. Every production system has a design capacity that defines it is maximum, sustainable level of production. No production can run at its design capacity forever; thus, there are periods of planned downtime for activities such as maintenance and product changeovers. The amount of time that the system is now producing products in is called the system's effective capacity level. During the effective capacity production period, there is downtime due to unplanned events, such as setup and adjustments of equipment and equipment failures. These unplanned time losses further reduce capacity to a gross capacity level that is the system's "Availability." During operations in the available capacity time, systems will not always be working at maximum speed due to planned speed reduction

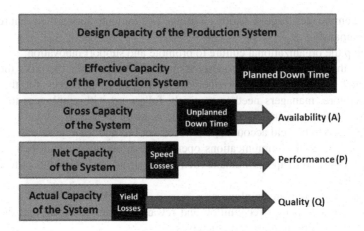

Fig. 10.1 Process capacity and OEE (Published with kind permission of © ABB, Inc. *Source* Wauters and Mathot [1]. All rights reserved)

and minor stoppages. These additional capacity losses affect the systems "Performance." Of the units produced, there is often a small percentage that does not conform to specifications; resulting in yield losses, which affect the system's "Quality."

The useful capacity level of the production system significantly affects the firm's productivity. The efficiency of the production system directly affects the systems output, and as such has a significant and direct effect on productivity. Productivity is not only affected by changes to the systems effectiveness, it is also affected by changes to the system's efficiency. This concept is illustrated in Fig. 10.2.

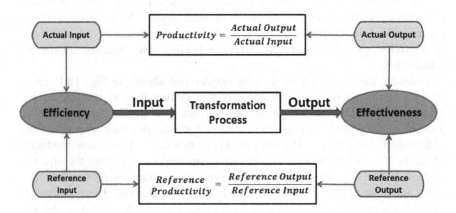

Fig. 10.2 Process productivity (Published with kind permission of © ABB, Inc. *Source* Wauters and Mathot [1]. All rights reserved)

The overall equipment effectiveness (OEE) metric is a measure of how effective the firm's production systems can transform inputs into outputs. The high-level formulation for OEE is given below:

$$\text{OEE} = \text{Availability} \times \text{Productivity} \times \text{Quality}$$

while the high-level formulation for total productivity is given as:

$$\text{Total Productivity} = \text{OEE} \times \text{Planning Factor}$$

The design capacity of this production system describes the maximum number of saleable units that can be produced if the system is operated at its maximum speed on a 24/7/365 schedule, given that everything is working correctly with no scrap losses and no time losses. In the real world, these conditions are highly unlikely. If mechanical equipment is continuously operating at full capacity, it will breakdown, usually with catastrophic results. To prevent the equipment from breaking and striving to maximize its useful life, planners will plan for maintenance shutdowns. The percentage of time that the production system is scheduled to be available versus its theoretical availability is defined as its planning factor (PF).

There are three principle causes of capacity losses in a production system: losses due to equipment malfunction, process loss, and external losses. Equipment malfunctions constitute a wide variety of causes, many of which are the result of either improper operations or poor maintenance. Process losses are caused by the type of use or treatment of work during production. External losses are related to circumstances that are outside of the control of the production and maintenance. The real value of the OEE metric is that, if equipment effectiveness needs to be improved, only the losses caused by machine malfunctions and process can be changed by the organization. OEE is a direct measure of those loss components. It therefore enables the firm to baseline the system and measure the amount of improvement due to changes in the system.

External losses are the result of either planned or unplanned events. Planned events typically include social events such as weekends and holidays, limited demand, system upgrades, and/or modifications. Unplanned events could include such activities as environmental events such as extreme weather, material shortages, and lack of personnel. Planned downtime affects the system's planning factor.

Unplanned downtime affects the processes' availability. The system's availability is measured as the percentage of time the equipment or operation was running compared to the available time. The most common reasons for downtime due to machine malfunctions are equipment failures and preventative maintenance. The most common reasons for downtime due to process are setup times, shift changes/daily maintenance, and change out of consumables (i.e., filters, cutting tools, and lubricants). For example, if a machine scheduled to be available to run for two shifts (16 h) but is only in production for 12 h as a result of setup time, breakdown or other down time, then its "Availability" is 75 % (12/16).

The "Performance" portion of OEE is related to the run speed of the process as compared to its maximum capability, called the rated speed. Common reasons for speed losses from machine malfunctions are as follows: technical imperfections that are correctable by the operator, process miscues that result in lost yield, shutdowns and startups triggered by maintenance or production requirements, and incorrect machine setting.

An example of performance loss can be seen as follows: if a machine produced 70 pieces per hour, but its rated capability is 100/h, then its "Performance" would be 70 % (70/100). Alternatively, the machine might be capable of producing 100 pieces per hour with the perfect part, but can only produce 90 units per hour on a few less perfect parts. If the machine's capability is baselined at 100 parts per hour regardless of which parts are run, the resulting measure would negatively impact the machine's OEE.

The "Quality" portion of OEE measures the number of good parts produced compared to the total number of parts started. The most common reasons for downtime due to machine malfunctions are given as: start-up and shutdowns related to maintenance or an incorrectly functioning machine. These types of quality losses occur because the process is yielding products that do not conform to the quality standards. The most common reasons for downtime due to process are process settings are not properly tuned to the quality standards, mistakes due to changeovers, or reduced speed setting to correct for deficient performance. For example, if 100 parts worth of materials and components were started in the process, but only 90 parts meet quality standards, the "Quality Factor" would be 90 % (90/100). Taking the three factor results from the process described, the OEE is given as follows:

$$OEE = 75\ \% \times 70\ \% \times 90\ \% = 47.25\ \%$$

In order to achieve an OEE of 100 %, it would require a machine to produce good quality every second of the available time at its top rated speed.

10.5.1 Using OEE to Drive Improvements

The key to use OEE as an improvement tool is in the analysis. Knowing that your OEE is 47.25 % does not help you understand that your process was unavailable 25 % of the time, or why it was unavailable for that much. To get the most usage from OEE, information must be captured to show where and why the losses

Table 10.2 Industry leading OEE thresholds (Published with kind permission of © ABB, Inc.

Industry	World class OEE (%)	Top industry performers OEE (%)
Manufacturing	85	60
Process	>90	>68
Metallurgy	75	55
Paper	95	>70
Cement	>80	60

Source Wauters and Mathot [1]. All rights reserved)

occurred. In other words, where are those buckets of "opportunities"? Knowing the complete breakdown of your OEE metric will make it apparent where the opportunities for improvement exist. To provide some points of reference wih various industries, Table 10.2 shows both industry leading OEE levels and world class levels.

10.6 Process Improvement Methodologies

To improve the effectiveness, efficiency, and productivity of a process, improvement activities need to focus on the reduction of manufacturing cycle times, the reduction in variability of outcomes, reduce costs, and upon the development of repeatable, measureable methodologies which generate outcomes that consistently conform to specification. These methodologies and tools include the following: Six sigma, Lean management, Lean six sigma, Re-engineering, Agile management, Just-In-Time, Quality circles, Single minute exchange of dies, Kaizen, Hoshin planning, Poka-Yoka, Design of experiments, Value analysis, and Process excellence. Many of these methodologies and tools are discussed extensively in other books. Following is a brief description of each.

10.6.1 Lean/Just-In-Time Management

Lean and Just-In-Time (JIT) are two management philosophies based upon the elimination of waste and the improvement of quality. As with TQM, both Lean and JIT are focused on the customer and on continuous improvements of the process. The principle difference between these two approaches is that JIT is focused on the production processes of the company, while Lean extends out across the supply chain and is focused on the business processes of the enterprise. In both of these philosophies, waste is defined as any none value adding activities, such as overproduction, waiting time, transportation, processing, and inventory.

10.6.1.1 Lean Thinking

Where six sigma is best applied to conformance problems; lean production is best at solving efficiency problems such as cycle times, cost, and throughput. The principle goal of the lean philosophy is to balance the flow of materials across the enterprise, thus, increasing its ability to produce a mix of products and services, while reducing the amount of resources used. To achieve this goal, lean focuses on identifying and eliminating non-value-adding activities throughout the enterprise; thus improving customer response times, reducing inventories, improving quality, and developing better human resources.

The building blocks (Fig. 10.3) of this goal are product design, process design, personnel and organizational elements, and manufacturing planning

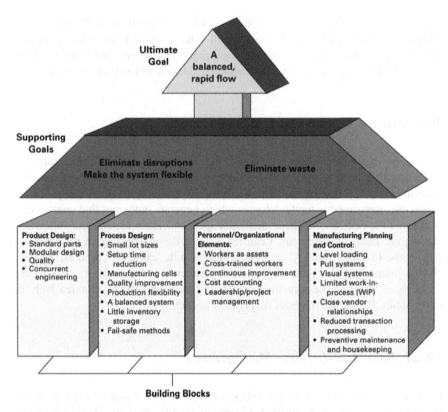

Fig. 10.3 The building blocks of lean (Published with kind permission of © McGraw-Hill Education. *Source* Vollman et al. [2]. All rights reserved)

and control. The first building block is product design. With the product design building block, lean promotes practices such as greater usage of standardized parts, modular design, quality, and concurrent engineering. Superior quality results do not just happen; they must be planned for and built into the products and services that the company offers its customers. Quality problems often create disruptions in the flow of materials through the production system. Any time the production is stopped, high costs are incurred, reducing the company's competitiveness; thus, it is imperative to avoid shutdowns and to quickly resolve the issues leading to a shutdown should they appear. Practices, such as standardized parts, lead to standardized work methods, which in turn develops workers that are very familiar with their jobs. Modular designs have the same affect. Both of these practices will reduce design and production costs. Designers must also understand the effects of quality-related specifications on the production systems and chose appropriate quality levels.

Another source of production shutdowns are engineering changes. Concurrent engineering practices can reduce these types of disruptions by improving the design planning process. Concurrent engineering is a product development approach in design engineering, manufacturing engineering, and other functions are integrated into a single team to reduce the elapsed time required to bring a new product to the market by parallelization of their respective tasks.

The process design building block is focused on the development of manufacturing practices that avoid the creation of scrap and reword. In this building block, there are eight practices that are critical to the success of lean production systems: small lot sizes, setup time reduction, manufacturing cells, quality improvement, production flexibility, a balanced production system, inventory reduction, and Poke-Yoke (fail-safe methods). The benefits associated with increased throughput are faster and more continuous material flows, resulting in reduced work-in-process inventories, greater scheduling flexibility, reduced carrying costs, reduced space requirements, and less clutter in the workplace.

One of the best methods for reducing costs and increasing the flexibility of production process is to reduce the time that production processes are offline (i.e., not producing products) due to changeovers and setups. When changeovers and setups times are high, not only are they expensive, but larger batch sizes are required in order to distribute the costs and kept the cost per unit manageable. Larger batch sizes reduce the flexibility of the production system in coping with changing customer demands; it also creates high level of inventory, which are themselves expensive.

When engaging in a SMED activity, the objective is to reduce the changeover/setup time to less than ten minutes; the lower the better. The starting point of SMED is to observe the current method from beginning to end. Once the main changeover/setup activities are identified, internal and external activities must be determined. Internal activities are those that must be performed on the equipment with the production process stopped. External activities are those which can be performed separately while the production process is operating. Next, streamline the changeover procedure. Frequently, changes to the equipment and the development of special tooling are necessary. Finally, train the operators in the new procedure.

Example: One of the processing activities for Part A is a milling operation, where three different cuts are made, each requiring a different cutting tool. After the part is positioned and clamped into place, the first cutting tools are locked into the toolholder, and the milling machine's turning speed and feed rate are adjusted, the first cut is made. To make the changeover to the second cutting tool, the machine is turned off, the first cutting tool removed. When the second cutting tool is locked into the toolholder, the turning speed and feed rates are readjusted, and the second cut is made. The process is repeated for the third cutting tool. The initial setup took approximately 45 min with each tooling change taking an additional 2 min. Total machine down time is 49 min per part.

Let us apply SMED. Instead of making three tooling changes, convert the mill to allow the usage of a fully articulated, multi-tool head that uses a quick-change chuck lock. This will reduce the initial loading of the cutting tools to be performed offline, and then take only about a minute to install when the machine is off. All subsequent tooling changes can be made with the machine running and take only a half-minute each. Next, design an interchangeable, clamping board that will hold parts during machining operations. The part is mounted to the board offline while the machine is operating. With the machine shutdown, the board is placed between a set of guide rails and locked into place with quarter-turn clamps. Part changeover time is reduced from 45 min of machine downtime to less than ten minutes.

Manufacturing cells allow for improved material flows while increasing equipment utilization and greater cross-training of operators. These work cells contain the machines and tools needed for processing parts that have similar processing requirements.

Disruptions and quality defects during the production effectively reduce the company's overall production capacity and as a result reduce profits and competitiveness. Therefore, problem solving is important; thus, bring us to the building block of quality improvement. There are two necessary components to problem solving and quality improvement. The first component is being able to detect problems when they occur. The second component is for someone to stop production and to correct the cause of the defects.

Given the overall goal of lean is to increase the firm's ability to produce a mix of products and services, while reducing the amount of resources used: An obstacle to this goal is bottlenecks in the process. Every system has a bottlenecking activity in it. The degree to which that bottlenecking operation affects material flows and throughput depends upon the inflexibilities in a systems design. Thus, it is important during the design of a production system to focus not only upon conformance to specifications but also flexibility. One mechanism for achieving this design goal is to balance the work load across the system by distributing the workload evenly among workstations. Another method is to devise alternate processing capabilities.

One of the principle sources of waste in the lean philosophy is excessive inventory. Inventories are necessary for buffering the production system against disruptions, but they also tend to cover up recurring problems that are never resolved. Instead of using inventory as a solution to work disruptions, focus on eliminating the causes of machine breakdowns and other types of disruptions.

Poka-Yoke is a Japanese term that means "mistake-proofing." Poka-Yoke is any mechanism that when applied in a manufacturing process helps employees avoid mistakes. Its primary purpose is to eliminate product defects by preventing, correcting, or drawing attention to human errors as they occur. It can be implemented using vision aids such as color codes gauges, or tool outlines in storage areas. It can also be implemented in product design. Examples of Poka-Yoke in products are the oversized grounding prong of a wall plug, the snap lock on telephone and computer network cords, and the internal alignment bar in UBS plugs. An example of Poka-Yoke in system designs is the switch in the car's gearshift that requires it to be in Park or Neutral before the car can be started.

There are three types of Poka-Yoke methods for detecting and preventing errors:

- The contact method which identifies product defects by testing the product's physical attributes.
- The fixed-value method which alerts the operator should be a set number of movements are not made.
- The motion-step method which determines whether the approved steps of the process have been followed.

Poka-Yoke can be implemented at any step of a manufacturing process. Poka-Yoke is a technique that can be used wherever an error can be made. There are several types of errors where this tool can be applied effectively:

- *Processing error*: Where an operation is missed or not performed according to standard operating procedure.
- *Setup error*: Where the wrong tooling is used or when setting a machine adjustment incorrectly.
- *Missing part*: Where one or more parts are not included in the assembly, welding, or other processes.
- *Improper part/item*: Where the wrong part is used in the process.
- *Operations error*: Occurs when an operation is carried out incorrectly.
- *Measurement error*: Occurs when an error is made in a machine adjustment, test measurement, or dimensions of a part coming in from a supplier.

There are five elements in the third building block of lean (personnel and organization): workers as assets, cross-trained workers, continuous improvement, cost accounting, and leadership/project management. These elements have either been discussed previously or are self-evident.

The fourth building block of lean is manufacturing planning and control, which includes: level loading, pull systems, visual systems, limited work-in-process inventories, close vender relationships, reduced transaction processing, and preventive maintenance and housekeeping.

10.6.1.2 Just-In-Time

JIT has been described as an inventory policy, as opposed to being a quality improvement policy. With the exception of establishing planned buffers inventory throughout the production system and linking the movement of this inventory to work authorization signals called "Kanbans," the JIT philosophy endorses a zero inventory approach, which is focused on having suppliers deliver materials and supplies as they are need each day. For this type of system to work properly, quality has to eliminate defects and schedules most be rigorously adhered to. Furthermore, employees must be given greater decision-making authority in the execution of their tasks. In addition, there should be a focus on the broadening of the task responsibilities. Thus, there are planning, scheduling, inventory, quality improvement, and organizational design elements inherent to the JIT philosophy.

10.6.2 Business Process Re-engineering

Business process re-engineering (BPR) was one of the first process improvement approaches to go beyond the continuous, incremental improvements focus of TQM, endorsing the big, dramatic improvements. This type of improvement approach often results in the total redesign of one or more processes within the business. BRP maintains all of the other elements of the TQM philosophy. Thus, it can be argued that BPR is just another tool in the TQM tool bag. There are several books out specifically dealing with BPR.

10.6.3 Agile Management

Agile manufacturing is a production/organizational philosophy that seek to increase competitive advantage by turning speed and agility into a core capability. The objective of this philosophy is to respond rapidly to customer needs and changes in the market. The key elements that enable this degree of flexibility are as follows: modular product designs, information technology, strong supplier relationships, and a knowledge-based organizational culture.

As a manufacturing philosophy designed to thrive in environments where continuous and unanticipated change is a norm, agile manufacturing systems require a set of resources that are often beyond the reach of most companies. Agile organizations have created processes, tools, and training that enable it to respond quickly to customer needs and market changes while still controlling costs and quality. Because of the necessity for a diverse set of resources, companies adopting this strategy must form partnership with other companies, where the group shares the resources and technologies of each in order to meet their objective. What makes this partnership work is the development of a manufacturing support technology that allows the companies, their designers and production personnel to share: (1) a common database of parts and products and (2) data on production capacities and problems.

10.6.4 Six Sigma

Six sigma is a highly flexible, structured set of techniques and tools for solving cross-functional problems. The objective of the six sigma philosophy is to maximize customer satisfaction by minimizing the defects (e.g., quality problems) that lead to customer dissatisfaction.

Utilizing a team-based approach along with quality management methods and tools, supported by a special infrastructure of people within the organization (e.g., Champions, Black Belts, Green Belts, and Yellow Belts) who are experts in the methods, six sigma projects are carried out following a defined sequence of steps and to achieve quantified value targets. The responsibility of a six sigma project team is to identify these inputs associated with the observed quality problem, remove them, and institute new controls to insure that the problem does not return. The core philosophy of six sigma is built around the following concepts:

- Think in terms of key business processes, customer requirements, and strategic objectives.
- Emphasizes quantifiable measures.
- Identify the critical-to-quality and key customer satisfaction metrics early in the process.
- Train and develop team members.
- Create highly qualified, subject matter experts who can support team efforts through the application of improvement tools and analysis.
- Set stretch goals for the improvement.

10.6.4.1 The Define Phase

The six sigma methodology has six phases, each a prerequisite to the next. The first phase is the defining phase. In this phase, the following factors must be identified and defined:

- A problem.
- The customer(s).
- Voice of the customer (VOC) and critical-to-quality (CTQs).
- The target process subject to DMAIC and other related business processes.
- Project targets or goal.
- Project boundaries or scope.
- A project charter is often created and agreed upon during the "Define step."

Define the Problem

Define the problem as observed by the customer and authorize a project team to investigate and solve it. Define the problem in operational terms that will facilitate further analysis. In other words, the deficiency to be corrected must be clearly articulated in observable and measureable terms, without assigning cause or blame.

Define the Business Case for the Project

Very few companies have unlimited finances and resources. As a result, they will typically have more needs that the means to satisfy them; thus, there must be a means for determining what projects will be supported and which ones will be either delayed or dismissed. This selection process typical involves a financial analysis, a risk assessment, and a logical argument for how the goals of the project support the strategic objectives of the company.

Selecting the Project

An important part of the defining phase is the selection of the project. In order to properly evaluate the potential of any improvement initiative, data are required. Some of the data needed during the evaluation process are as follows: The sources of the complaints, the competitions capabilities with respect to your firm, the costs associated with the problem, and information on the deficiencies of your own processes, including their effect on employee morale. With this level of data, you should be able to determine which problems are the most important and have the greatest future impacts on competitiveness and customer satisfaction.

When selecting the improvement project, you should focus upon those chronic issues that are having significant impact on customer satisfaction or costs or will significantly enhance customer satisfaction and/or employee morale, reduce costs, or provide a return on the investment.

The project should not take longer than one year to complete. If it does, than determine if it can be subdivided into small projects. In addition to understanding the duration of the project, what are the expected risks to a successful completion?

Charter the Project

Once executive management has approved the project, a project charter is created authorizing the work. This charter defines the project, its objectives, the deliverables, the project team and its sponsor, the customer, the critical-to-quality metrics, the timeline, and the resources needed for the project. In essences, the project charter is a contract between the project team and management, providing the team authority to draw on company funds and utilizes its resources.

10.6.4.2 The Measure Phase

In the second phase, the controllable variables of the problem are identified and data are collected. In this phase, the process is focused on developing an understanding of the processes performance and in collecting data necessary for analysis. By viewing the relationship between process performance and customer satisfaction as a mathematic function, the relationship is expressed as:

$$Y = f(X)$$

where Y is the set of CTQ's, and X represents the set of critical input variables that affect Y. The key to successfully solving this problem is to determine which X is within the company's control. The following considerations can help clarify the data collection effort:

1. Formulate questions related to the specific needs of the project.
2. Select and use the appropriate analysis tools.
3. Identify and collect the necessary data.
4. Define the each piece of data to be collected and the appropriate collection method in operational terms.
5. Select unbiased collectors and train them in the collection method process. The training should be focused on what data are to be collected, and insuring that the operators technique is repeatable.
6. Design the data collection forms to be simple and informative; thus, insuring they are filled out correctly.
7. Test the data collection methods and forms, to insure that the collection methodology is repeatable and accurate.
8. Audit the data collection process and validate the results.

10.6.4.3 The Analyze Phase

The third phase of six sigma involves the posing of theories on the root causes of the problem, the testing of those theories, and the identification of root causes to the problem. The analysis starts with developing an understanding of the process. There are numerous tools available supporting this activity, ranging from flow charts, to cause and effect diagrams, to value stream mapping.

The next step is to understand why defects, errors, or excessive variance is occurring in the process. The goal is to ascertain the fundamental causes (root cause) of problems. There is a variety of reasons for process error: out-of-spec materials, faulty machinery, deficient work methods, human error, and/or measurement errors.

It is important that before any corrections begin, clear proof of the cause has been established. The best method of achieving this is to formulate theories as to the cause and test those theories using the collected data and rigorous analytical techniques.

10.6.4.4 The Improve Phase

In the improvement phase of the six sigma methodology, alternative approaches to the removable of the identified root causes of the problem are generated. The solution alternatives are developed, analyzed, and the best one selected and implemented. The improvement should not only remove the root cause, it should produce optimal results for both the customer and the organization.

10.6.4.5 The Control Phase

With the solution in place, the fifth phase designs and implements a new control system that will prevent the original problem from returning. No improvement is complete unless there is a procedure in place that explicitly explains how to hold the gains. Otherwise, the benefits from the improvement project will be lost overtime.

10.6.4.6 What Makes Six Sigma Work?

Though six sigma and TQM have many similarities, a few differences allow six sigma to stand-alone and make it work better than any other quality methodologies. The principle difference lies in six sigma's global perspective. Six points of difference can explain this global perspective.

1. *Top Management Support*: The use of universal cost orient metrics and results that provide new levels of competition make it easier for managers to support.
2. *Team-based Problem Solving*: Six sigma espouses the development of highly trained specialists to support and lead improvement teams: Executive Champion, Deployment Champions, Project Champions, Master Black Belts, Black Belts, and Green Belts.
3. *Training*: There is greater emphasis on training everyone in six sigma and analytical techniques.
4. *Metrics*: Six sigma not only focuses on customer CTQ, but also measures process performance at a higher standard called Defects per Million Opportunities (DPMO). The use of this metric provides greater opportunities for comparisons across divisions, departments, and processes within the company.
5. *Use of Improvement Teams*: Six sigma demands that projects involve highly trained, cross-functional, and empowered teams to locate and make improvements.
6. *Corporate Attitude and Culture*: Just the implementation of six sigma generates an environment that encourages the creation of continuous improvement effects.

10.6.5 Quality Circles

A quality circle is a group of employees (volunteers) who meet on a regular basis to solve problems affecting their work area. These circles typically consist of between 3 and 12 members with a facilitator. Essentially, a quality circle is a participative management system where workers can make suggestions and improvements that improve the organization.

Quality circles focus on problems associated with quality, productivity, safety, job structure, process flow, control mechanisms, and esthetics of the work area.

Examples of the ways that quality circles improve the company and indirectly the employees are as follows:

- To improve the quality and productivity.
- To reduce the cost of products or services by reducing unnecessary errors and defects, improving safety, and increasing the effective utilization of resources.
- To identify and solve work-related problems.
- To tap the creative intelligence of employees.
- To engage employees to develop and use their knowledge and skills and then applying them to a wide range of challenging tasks.
- To improve communication within the company.
- To provide avenues for recognition, achievement, and self-development.

10.6.6 Kaizen

Kaizen is a Sino-Japanese word, which simply means "good change," with no other meanings espousing concepts such as "continuous" or "philosophy." Kaizen simply refers to any improvement, either one-time or continuous, large or small, that increased the quality, or effectiveness, or efficiency of a process. However, in practice, it typically applies to measures for implementing continuous improvement, or even taken to mean a "Japanese philosophy" thereof.

Kaizen is a process that when applied to daily activities, goes beyond simple productivity improvement. Its purpose is to humanize the workplace by eliminating overly difficult tasks by teaching employees how to perform experiments on their work using the scientific method and how to spot and eliminate waste. The actual application of Kaizen can be through a suggestion system, either at the individual level, or at the small group level, or at the large group level. As with most all quality initiatives, the successful implementation of Kaizen requires the participation of people at all levels of the organization, from the CEO down to the janitorial staff, as well as external stakeholders when applicable.

Kaizen groups are often guided through the kaizen process by a line supervisor. When kaizen is applied on a broad, cross-departmental scale, it generates total quality management and frees human efforts through improving productivity and the utilization of machines and computing power. While kaizen often delivers small improvements, the culture of continually aligning small improvements and implementing standardization that is developed throughout the company frequently yields large results in the form of compound productivity improvement.

10.6.7 Design of Experiments

In any system, there will be variability. This variability makes outcomes less predictable. The primary purpose of any improvement activity is to reduce and/ or eliminate variability. In a system, there are several factors that control the functionality and quality of its output. These factors often have several levels (called treatments) at which they interact with the system. Some of these factors are controllable by the operator, while some are not. To understand how the controllable factors influence the outcomes from the system, a series of experiments need to be conducted. The problem is that experiments tend to be time consuming and usually are expensive. Thus, running a full spectrum of experiences for every level of each factor is not feasible.

In design of experiments (DOE), a subset of factors and treatments are selected that best represent the spectrum of possible. Formal experiments are conducted to determine the outcomes. There are several methods of designing the actual experiments in order to improve the predictability of results. These methods include: comparison, randomization, replication, blocking, orthogonal, and factorial experiments. In designing the experiment, the following considerations need to be addressed:

1. How many factors does the design have and are the levels of these factors fixed or random?
2. Are control conditions needed? If so, what should they be?
3. Manipulation checks: Did the manipulation really work?
4. What are the background variables?
5. What sample size is needed to insure generalizability and power?
6. What is the relevance of interactions between factors?
7. What is the influence of delayed effects of substantive factors on outcomes?
8. How do response shifts affect self-report measures?
9. How feasible are repeated applications of the same measurement instruments to the same units at different occasions, with a post-test and follow-up tests?
10. Are there latent variables?
11. What is the feasibility of subsequent application of different conditions to the same units?
12. How many treatments of each control factors need to be taken into account?

At the conclusion of the formal testing, the data is statistically analyzed to determine the degree of affect and significance of each treatment on the results. It is best that the process being studied is in reasonable statistical control prior to conducting the designed experiments. When process stability is not possible, design methods such as blocking, replication, and randomization can be used as controls to increase the power of the results. It is important that uncontrolled influences do not skew the findings of the study. Additionally, it is important that the effects of spurious, intervening, and antecedent variables are eliminated from the experiment.

References

1. Wauters, F., & Mathot, J. (2002). *OEE overall equipment effectiveness*. White paper. Zurich: ABB Inc.
2. Vollman, T. E., Berry, W. L., & Whybark, D. C. (2005). *Manufacturing planning and control systems* (5th ed.). New York, NY: Irwin/McGraw Hill Companies Inc.

Chapter 11
The Dimensions of Supply Chain Quality

With the increased level of competition due to globalization and improved telecommunication technologies, many of the natural barriers to enter for many markets became inconsequential. One of the most significant effects of increased competition is the erosion of profit margins. As such, companies need to find new ways of improving cost efficiencies: The answer was to better manage the supply chain. A supply chain consists of those interconnected companies, both upstream (i.e., supply channels) and downstream (i.e., distribution channels), and the ultimate consumer.

Supply chain management (SCM) began as a field of academic study and industry practice in the middle 1980s, in response to needs from both logistics management and marketing. Logistics management had a need for better synchronization of operations across both functional and organization boundaries. Marketing on the other hand recognized the need for improved communications between organizations involved in the development, production, and delivery of products and services, particularly the communications of customer requirements. SCM has been defined as "the systemic, strategic coordination of the traditional business functions and the tactics across these business functions within a particular company and across businesses within the supply chain, for the purposes of improving the long-term performance of the individual companies and the supply chain as a whole" [1]. SCM includes the management of supply and demand, the sourcing of materials and services, manufacturing, assembly, inventory management, logistics, warehousing, management of distribution channels, and delivery to customers.

During the 1970s and 1980s, innovations in technology and management philosophies with respect to quality, processes, and production practices improved the performance of the firm's operations significantly. In addition, product quality could no longer provide the competitive advantage that it did prior to these improvements in

© Springer-Verlag London 2015   141
G.N. Kenyon and K.C. Sen, *The Perception of Quality*,
DOI 10.1007/978-1-4471-6627-6_11

management philosophy and practices, and in markets in general. The customers today views product quality as merely a qualifying factor and value as the order winner. As a result, management has started thinking about business processes, logistics, and supply and distribution channels as an enterprise. Where previously the focus was placed on suppliers for lower costs and the optimization of internal manufacturing processes, now it is on creating a partnership with supply chain participants and finding ways of reducing overall cost structures while improving value to the customer.

With improvements in production capabilities, speed became a competitive issue. With this increased focus on speed, logistics and distribution channels came under management scrutiny. The knowledge and methodologies developed from this increased attention gave birth to the field of study and practice called SCM. The benefits of an effective SCM program often increase with the growth of competition in the marketplace.

Due to the rapidly changing business environment characterizing the early 1980s, the philosophy of SCM emerged into being about holistic decision making across business activates, processes, supply lanes, and distribution channels. If a company wished to maximize its profitability and competitiveness, it could no longer view its operations' function in isolation. The operations' function of a business is both a customer for other businesses and a supplier to its own customers. Over time, the study and practice of SCM has extended the concept of functional integration and the coordination of business activities beyond traditional organizational boundaries to include all firms in the supply chain, such that each member (firm) of a supply chain contributes to the improvement of the chain's competitiveness.

This network of firms contains the resources and processes that together produce products and services for the marketplace. The objective of SCM is the maximize competitiveness and profitability of the whole supply chain. The goal of SCM is to meet the company's customer service objectives while minimizing inventory and related cost. Furthermore, SCM is focused on the interactions of supply chain participants to produce a product or service that will provide the best comparative value for the end user.

By the 1990s, modern advances in communications technologies and transportation methods have globalized everyone's perceptive. With ready access to information about products, technology, and ideas worldwide, the customer's knowledge frequently exceeded the firm's. As a result, customers were demanding more from companies because they know that if the chosen company does not perform, its competitor will. Flexibility became a major competitive strategy, and large multinational companies were competing internationally through their supply chains. At this level of competition, managers have to know more about how the activities and decisions of their suppliers, operations, logistics, subcontractors, strategic partners, and channel partners affected business performance. By focusing their skills in total quality management, high-performance operations, and innovations in communication and technology, managers were able to improve their supply chain performance. Efforts to balance customer demands with profitability have primarily been focused upon flexible organizations, organizational relationships, supply chain coordination, communications, outsourcing of non-core competencies, built-to-order manufacturing strategies, inventory management, and cost control.

The objective of SCM is to maximize competitiveness and profitability of the whole supply chain network while simultaneously meeting the company's customer service objectives and minimizing inventories and operational costs. To achieve this objective, SCM practitioners need to focus on the interactions between supply chain participants as they produce a product or service that will provide the best comparative value for the end user. Business activities that fall under the SCM umbrella are order management, procurement, production, distribution, customer service, customer relationship management (CRM), supplier relationship management (SRM), and partner relationship management.

Dr. Marshall Fisher, Professor of Operations and Information Management at the University of Pennsylvania's Wharton School of Business, speculates that the inefficiencies typically found in most companies are the result of a mismatch between the type of products that a company was producing and the type of supply chain it has created. He describes the two basic types of supply chains: physically efficient supply chains and market-responsive supply chains. The dimensions by which supply chains are created and managed are the following: the structure of the supply chain, the relationships between the supply chain partners, and its material and information flows.

11.1 The Integration Dimension

SCI is typically characterized by the degree to which the following elements are successfully implemented: relationship management (RM), synchronization, collaboration, and information technology. There are two levels of SCI: internal and external. Furthermore, there are two distinct areas of focus with external SCI: upstream toward suppliers or downstream toward customers. The effect of SCI on competitive capabilities such as product quality, delivery, cost, customer service, marketing, differentiation, reliability, process flexibility, and new product development has been extensively researched.

Several studies of the automotive industry have found a positive relationship between internal and external SCI, as well as, SCI enhanced product quality, costs, and delivery capabilities. Another study, this time using data from US manufacturers, found that customer and supplier integration activities provide benefits in business performance. With respect to strategic supplier integration, significantly market performance benefits were found, but no significant effects related to customer satisfaction.

SCI affects product quality in two ways: conformance to specifications and design. The principle mechanisms by which SCI promotes performance benefits are through building trust, commitment, adaptation, communication, and collaboration. In a study that examined the relationship between these factors and product quality, it was found that SCI activities did improve product quality and improvements in quality lead to increased customer satisfaction.

11.1.1 Relationship Management (RM)

To be successful, RM must be approached with a long-term perspective which is geared toward building trust and communication for the purpose and engendering cooperation between multiple firms for mutual economic gains. Additional benefits include the sharing of important information and improved knowledge management. In addition, long-term relationships' material and information flow are improved, bottlenecks eliminated, and delivery flexibility increased.

11.1.1.1 Customer Relationship Management (CRM)

CRM is a management practice designed to manage, improve, and facilitate sales, support, and related interactions with customers, prospective customers, and business partners (distribution channels). The principle characteristics typically found in CRM are as follows: a customer-centric, response-based service; sales force automation; use of technology; and opportunity-sensitive management focus for addressing unexpected growth and demand. Best practices associated with CRM are as follows: ability to identify potential customer and effectively and timely communicate; ability to reward desired customer behaviors; and ability to recognize key customers, two-way interaction with customers, collaboration with customers, and a systemic method for enabling customer advocacy.

11.1.1.2 Supplier Relationship Management (SRM)

SRM starts with strategic planning. The primary focus of SRM is the managing of all third-party interactions involving the sourcing and supply of products and/or services, with the objective of maximizing the value of those interactions. Similar to CRM, the goal of SRM is to develop a mutually beneficial, two-way relationship with suppliers so as to increase competitive advantage.

Best practices associated with SRM are the following: fostering trust, cross-functional collaboration, synchronization, communications, a clear and jointly agreed governance framework, value analysis, and enterprise-wide technology and systems. Other facilitating practices include segmentation, supplier summits, joint strategic business planning meetings, and joint operational business reviews.

11.1.1.3 Concurrent Engineering (CE)

One practice associated with customer and supplier interactions is CE. Whether one is developing a new product, service, or process, or if one is trying to improve an existing product, service, or process, CE can increase the quality of the result as well as lower the cost and shorten the time frame of execution. In addition, CE can improve the responsiveness of an organization in dealing with changes in its business environment. CE accomplishes these benefits by involving all relevant stakeholders of an activity early on in the process and by promoting the sharing of important information. By involving these key stakeholders early, their insights on how and why various actions can or should be taken are often invaluable for several reasons: These stakeholders bring knowledge from different perspectives and backgrounds; suppliers, on the other hand, have greater knowledge of the usages and applications of the materials and components they are providing the project; and customers best understand what they want, how they are going to use the results of the project, the limitations of their own processes and capabilities, and the end result they are expecting.

An example of CE can be seen in how The Boeing Company designed its 777 long-range, wide-body, twin-engine, jet airliners. A cross-functional design team was assembled for each of the major components of the plane. One such team was responsible for the design of the wing flaps. The members of this design team included design and manufacturing engineers, purchasing, inventory management, and numerous other key stakeholders, including a customer representative from United Airline's (UA) Field Servicing and Maintenance. During the design conceptualization phase, the UA representative questioned the logic of a single-piece composite structure for the flap. His concern was that during ground handling, wing flaps could become damaged. When this happens, the flap must be removed from the plane and repaired. The repair process involves cutting out the damaged area and patching the hole. The flap must then be placed in an autoclave where high-temperature and pressure can be applied to cure the composite patch. Given that the wing flap is 66 feet in length and that UA's field maintenance autoclaves can only handle parts up to 35 feet in length, UA would have to ship the damaged flaps back to Boeing for repair. This would necessitate the grounding of planes for several weeks. The carrying of spare flaps in inventory would alleviate the grounding of the plane, but the additional field inventory would incur a significant cost. Both alternatives would increase the operating cost of the planes beyond what was economically feasible. As a result, the designs were changed, cutting the flap into two sections, thus accommodating UA's needs and cost constraints, without any loss of functionality or addition costs to production. If this issue had gone unnoticed till customer checkout, it would have caused the cancellation of UA's orders and most likely the overall project.

11.1.1.4 Synchronization

Consumer value is created not by individual firms but rather by business enterprises consisting of multiple interdependent companies working together to create the value the customer desires. The competitiveness of an enterprise is dependent on both the competitiveness of the individual firms and the nature of the linkages between the firms in the supply chain. To maximize value of a supply chain, the coordination mechanisms used to manage its member organization's capabilities need to be matched appropriately to the market structure they are serving. Coordination mechanisms are those practices and tools that enable better communication of information, facilitate continuous improvement, speed up response time, improve product quality, and create closer relationships between customers and suppliers.

Vertical partnerships are defined as an arrangement between a buyer and seller to facilitate a mutually satisfying exchange over time, which leaves the operation and control of the two businesses substantially independent. There are four key aspects to vertical partnerships: Partnerships are entered into freely, partnerships must offer mutual benefits, these benefits occur over time, and partners remain substantially independent. There are five benefits to vertical partnerships: improved market access, improved communications, higher profit margins, greater discipline, and higher barriers to entry. Resistance from key stakeholders can often be explained by their refusal to change old ways of doing things, or by their inability to perceive the benefits of a proposed change. The degree to which supply chain activities can be coordinated is determined by the trade-off between the cost of coordination and the cost of inefficiencies within the system.

11.1.1.5 Collaboration

When all functional departments, from the CEO through accounting and invoicing to materials handling, are working together and sharing the same information, companies can increase their ability to meet its strategic goals seamlessly. Externally focused collaboration seeks to achieve the same benefits across the entire supply chain with SCM. Through the usage of advanced information technologies, systems integration had become the facilitating thread that ties all these pieces of equipment, software, and business processes together. The more complex a system is, the greater the computational and organizational burdens, and the greater the risk of system breakdown. This risk can be reduced through a well-designed, well-integrated, robust system.

11.1.1.6 Technology Management

Changes in supply chain operations often included applications of information technology (IT). IT-based solutions for improving supply chain performance will reduce goal incongruence of SC partners by making more variables observable and, hence, reducing actions that negatively impact on the principal's welfare, by disseminating private information broadly in the supply chain and thereby reducing problems associated with adverse selection and local knowledge. It offers the opportunity to improve customer relationships and strengthen customer loyalty. Other benefits espoused by IT advocates with respect to e-commerce are the following:

- It improves the efficiency of current relationships.
- It facilitates the hooking-up with new buyers and sellers.
- It provides a means of mutual communication and interaction with consumers.

- It improves the efficiency of information transfer.
- It provides a clearinghouse for information.
- It can lower transaction costs.
- It can help employees be more effective.
- It can improve the coordination between supply chain partners.

When making operational changes to the supply chain, consideration must be given to the redesigning of control mechanisms, such as incentives. Based upon the notion of rational self-interested agents, the principal seeks to influence the agents' behavior by designing incentives, information, and decision-making authority in such a way that, while maximizing his or her welfare, the agent will, to the extent possible, also maximize the principal's welfare. Incentive issues will arise when either the actions of an individual cannot be observed or when one individual has information not known or possessed by the other individual.

By leveraging the cost-cutting benefits typically available with IT through the streamlining of activities and the utilization of this technology to create electronic communities (networks), closer relationships can be fostered among supply chain members. With the creation of electronic communities, traditional barriers that have existed between supply chain participants could be lowered. By using a multitiered structure that provides different vertical and horizontal interactions, electronic communities can increase an organization's exposure, thus allowing a more orderly and efficient process. The potential results would be to lower the cost structure across an entire industry such that all participants would benefit.

It is generally agreed that IT reduces the cost of coordination, and the lack of coordination typically results in the supply chain holding inefficiencies in the form of inventory buffers, underutilized capacity, lost sales, etc. IT in and of itself is no guaranty of success. IT can only increase productivity when it is properly applied to enable improvements in the way a company does business. A holistic approach that looks at all of a company's processes must be utilized.

11.2 The Process Management Dimension

Supply chains are increasingly becoming more global and interconnected. Companies continue to struggle to reduce their exposure to shocks and disruptions that can damage their competitive advantage. To achieve this goal, SCM emphasizes cross-functional links and seeks to manage hose linkages. There are eight business principle processes that comprise the focus of SCM. They are CRM, customer service management, demand management, order fulfillment, manufacturing flow management, procurement, product development and commercialization, and returns management. These eight processes span the length of the supply chain, as well as cut across functional silos within each firm. Figure 11.1 provides a diagram of the relationship of these eight key business processes and the supply chain. Though these processes affect the dimensions of SCM differently and

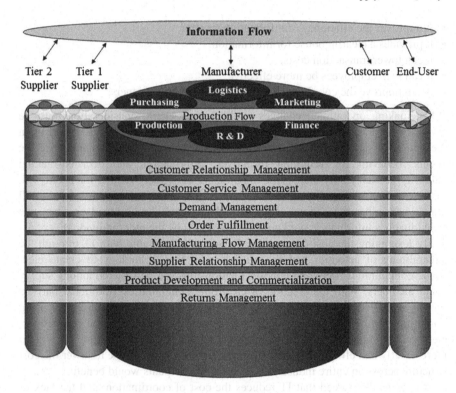

Fig. 11.1 Supply chain management processes (Published with kind permission of © Emerald Group Publishing. *Source* Cooper et al. [4]. All rights reserved)

in some cases affect multiple dimensions simultaneously, they must be managed effectively in order for the firm and its enterprise partners to meet their strategic objectives.

A survey of 600 global manufacturing companies made by Deloitte looked at the supply chain operations of these companies and the increasing complexity of managing the supply chain. From this survey, it was found that about 7 % of the companies had successfully handled the increasing complexity of their supply chains. These few companies (the 7 %) were able to maintain profit margins that were 73 % greater than manufacturers with simpler supply chains. In contrasting the success to the few against the many, this study found six significant factors in the management of supply chain complexity and overall profitability. These factors are as follows:

- Relentless pursuit of synchronization
- A holistic view of the business
- Profitability from the entire product life cycle
- Collaboration with customers
- Investment to technology
- Flexibility

Another study conducted by Accenture, INSEAD, and Stanford University analyzed 636 global companies, and four key findings were made. First, there was a strong connection between superior supply chain performance and financial success. Second, leading companies incorporated supply chains into their business strategies and devoted significant attention to designing integrated operating models. Third, leading supply chain companies build innovation into their operating models. The fourth key finding was that leading supply chain companies rigorously executed against their strategies and capabilities and constantly adapted them to changing market demands.

11.3   The Product Life Cycle Management (PLM) Dimension

PLM is the process of managing a product or service from its inception through to the end of its life cycle. As a process, PLM integrates human resource, data, processes, and business systems, while providing information to stakeholders. The objective of PLM is to maximize the profit potential of a product during its life cycle, while minimizing adverse effects to the organization. A product's life cycle has four distinct stages, each with their own characteristics. The basic management implications associated with each stage provide the firm with different challenges. The benefits typically realized from PLM are the following:

- Reduced time to market.
- Increased full price sales.
- Improved quality and reliability.
- Ability to quickly identify potential sales opportunities and other revenue contributions.
- Reduced waste.
- Reduced R&D and engineering costs.
- Improved forecasting.
- Maximized supply chain collaboration.

11.3.1   Stages of the Product Life cycle

Products have four distinctive stages in their life cycle: the introduction stage, the growth stage, the mature stage, and the decline stage.

11.3.1.1   Introduction Stage

The introduction stage is initiated with concept definition of the product or service. It is during this stage that the market potential of the product is determined alone with customer requirements and expectations. After the concept's feasibility has been determined, the product is researched, developed, planned, designed, and tested. Additionally, the manufacturing processes and methodologies are designed, and procurement activities are started. Finally, the product is market tested and, if successful, is placed into full production.

11.3.1.2 Growth Stage

The growth stage is typically characterized by a strong growth in sales and profits. Based upon forecasts of the expected rate of growth, production capabilities are expanded. Typically, companies will engage in promotional activities to maximize the potential of this growth stage during this stage.

11.3.1.3 Maturity Stage

The maturity stage is characterized by a stable customer demand pattern. The product has been fully established, and the objective for the manufacturer is to maintain the established market share. Product modifications and/or process improvements are made only if analysis confirms an increase in competitive advantage.

11.3.1.4 Decline Stage

Eventually, customer demands for the product will become sated or a next generate product will be introduced, and the market for a product will start to shrink. Though this decline may be inevitable, there may still exist opportunities for companies to make some profit by switching to less expensive production methods and cheaper markets. Ultimately, the company must plan and manage the obsolescence of the product and its discontinuation.

11.3.2 The Logistics Dimension

Logistics is a critical part of the supply chain and, though a separate functional activity, is often integrally linked with inventory management and/or SCM. The Council of SCM Professionals define logistics as being the "part of SCM that plans, implements, and controls the efficient, effective forward and reverse flow and storage of goods, services, and related information between the point of origin and the point of consumption in order to meet customers' requirements. Logistics management is an integrating function which coordinates and optimizes all logistics activities, as well as integrates logistics activities with other functions, including marketing, sales, manufacturing, finance, and information technology" [2].

The interfirm activities associated with logistics include the management of order processing, inventory management, transportation, warehousing, material handling, and packaging. The intrafirm activities associated with logistics encompass the coordination and synchronization of materials flow and information across the supply chain.

Logistics managers are typically responsible for designing and overseeing the systems that are capable of controlling the movement of raw materials, work-in-process, and finished inventories, as well as providing timely information of the location, quantity, condition, and availability of these inventories, at the lowest total cost. To a degree, these responsibilities overlap those of operations managers (i.e., manufacturing and inventory). Ultimately, the success of a logistics system is measured by its ability to deliver a customer's completed order when, where, and how they have specified it. When done effectively, logistics will contribute significantly to the firm's competitive advantage.

Logistics impacts the firms' competitive positioning in two ways. First, it is a necessary function for connecting production with consumption; thus, it is a cost center. Second, as a service, it can provide significant competitive advantage through differentiation, thus making it an integral part of quality and a factor in generating customer satisfaction and loyalty.

As an aspect of quality, logistics can be viewed from two basic perspectives: objectively and subjectively. Form the objective perspective, logistical activities are considered physical objects that can be observed and are possessing attributes that can be evaluated. From the subjective perspective, logistics is a service whose quality can only be evaluated by the customer.

Providing value in logistics entails ensuring an appropriate level of service at the lowest possible cost. The level of service offered in any logistics system is driven by two components: inventory availability and operational performance. With the inventory component, there are three basic functions performed: buffering the transformational processes against unexpected events such as excess customer demands or production failures, meaning planning customer satisfaction levels, and hedging against inflationary pressures in the marketplace. As an operational component, logistics not only provides the seamless flow of materials through the production system, but also is the last contact point for the firm with its customer.

Another important consideration in logistics quality is strategic segmentation. Different customers have different logistical requirements, and so goals are best set on a case-by-case basis. One customer may be best served with slow delivery at a low cost, while another requires overnight service, at a high cost. Accordingly, logistics quality should also account for segmentations' difference by basing performance against a target, where the quality of the service is matched on a customer by customer basis.

Research has identified eight attributes that define the quality of a logistics system: timeliness, completeness, information quality, availability, flexibility, contact quality, returns, and cost. For the average consumer, the dimensions of timeliness, flexibility, and reliability carry the most weight. In the case of the commercial customer, there is significantly more emphasis on accuracy and timeliness.

11.3.3 The Timeliness Attribute

Timeliness is a function of speed and reliability. When a supplier delivers an order on-time, it facilitates his commercial customer's ability to streamline their inventories. This reduces their operating cost and creating value. It also avoids dissatisfaction with both the commercial customer and the consumer.

Suppliers that have the ability to adjust their deliveries based upon changes that affect the customer, such as spikes in demand or changes in the product mix being demanded, enhance their customers' ability to adjust to their customers' needs. The reliability aspect of timeliness involves consistency in meeting delivery time and location specifications.

11.3.4 The Completeness Attribute

Completeness is a function of accuracy, condition, and perceived quality. The attribute of accuracy can be measured by how closely the items shipped match-up with the customer's order. If an item is missing from an order, the customer will be unable to satisfy their own needs. Thus, they will be dissatisfied, and many never consider using that supplier again. The attribute of condition refers to the lack of damage to the items received. If an item in the order is damaged, the customer will perceive it as being worse than it had not included, but now they must experience the aggravation of having to return the item and getting it replaced. The attribute of perceived quality refers to how well the items ordered perform the functions for which they were designed, as well as the degree to which that performance meets the customer's needs and expectations.

The average consumer is relatively more relaxed about the timeliness and completeness of their orders than the average commercial customer, as long as the requested items arrive relatively close together close to the requested time frame. Because of the needs of their own customers, commercial customers place significantly more emphasis on order accuracy and timeliness.

11.3.5 The Information Attribute

Information quality is related to the customer's perception of how informative the information provided regarding the products and services that they had ordered. For information to be useful to the customer in decision making, it must be of sufficient quality and readily available.

Technology has had a revolutionary effect on the quality of logistical services. One of the areas where technology has had a major effect is in supplier and CRM. By incorporating information technologies into the enterprise structure, companies have facilitated information gathering, processing, and distribution. This should improve decision making, both internally and across the supply chain. One of the main advantages in the incorporation of information technologies in the logistics structure is that it does not need to mirror the physical flow of materials; thus, it can be used to shorten the channel and reduce the number of intermediaries, to directly connect with suppliers and customers. This ability to separate the information flow from the physical flow of materials allows the firm to improve the speed of communications, reduce information related transaction costs, and provide the ability to optimize the information flows and physical flows separately, thus facilitating productivity.

11.3.6 The Availability Attribute

Though product availability is dependent upon the firm's production capabilities and inventory management practices, the logistics of delivering customer orders is largely a scheduling issue. Dependent upon the availability of current and anticipated inventories, the company has the opportunity to increase the value-add to customers by ascertaining their specific needs with respect to when, where, and how much of specific order items are needed and scheduling deliveries that best suit those needs.

Inventory availability is widely measured using stock out measures. Logistical performance is typically assessed using performance measures such as speed of delivery, consistency in delivery across all shipments, and time to recover if a failure does occur. Research has found that the operational elements of logistics can have a significant effect on product availability, condition, delivery reliability, and speed.

11.3.7 The Flexibility Attribute

Flexibility in logistics implies a highly differentiated value-adding process that is driven primarily by market and customer demands. The greater the flexibility of a logistics system, the greater its ability to adjust to changes in customer orders, such as quantities, location, and/or time. Various practices such as delayed differentiation have been used to increase the flexibility of distribution systems. Technology, such as enterprise resource planning systems, will enhance a system's flexibility by providing timely information.

11.3.8 The Service Experience Attribute

Logistics is both an operational aspect of the company and a service offered to the customer. As an operational element, poor performance will impede the firm's ability to produce products or services in a timely manner, thus causing the firm to either miss delivery dates or assume additional inventory costs by carrying large than necessary safety stocks throughout the production system. As a service element, logistic personnel are the firm's last point of face-to-face contact with the customer. Therefore, logistical personnel will have a significant influence on the customer's final perception of the overall transaction.

The customer's perception of the quality of logistics services is a function of both outcomes and process. In other words, how the task is performed is as important as the task itself. With respect to measuring the quality of a logistics interaction, the customer will focus on the professionalism and skill of the personnel, their attitude, behavior, and perceived trustworthiness, as well as the accessibility, reliability, and flexibility of the service process, and the reputation of the service provider. As with the perceived quality of a service, delivery quality is based upon the difference between what is expected and what was experienced. Thus, ensuring that delivery personnel are properly trained is important.

11.3.9 The Warranties and Returns Attribute

The competitive goal of every company is to satisfy their customer. The expectation associated with this goal is that by satisfying customers, the company can build loyalty within that customer group and attract new customers. Through the goods and services that they sell, companies are seeking to impress their customer with their understanding of the customer's needs and desires, along with their creativity in meeting those needs. However, when those products or services fail to meet the customer's expectations or needs, everything the company sought to accomplish is at risk. Their last opportunity to recover from this failure and to sway the customer to give them another change is based on how they handle the returns and/or warranties. Furthermore, how the company handles the return of failed products can influence their cost structures and profitability.

There are numerous reasons why customers return products: They were damaged, broken, and of poor quality or did not work properly; the products received were unordered, wrong, out of date, or not the correct quantity. By the very nature of returning a product, customers are going through a negative experience that can significantly affect their satisfaction and the company's credibility. Thus, by establishing a returns process that is customer friendly can transform a negative experience into a positive for both the customer and the company.

With respect to costs, it has been estimated that for the average company, reverse logistics costs average more than 9 % of total logistics costs. In fact, the Reverse Logistics Executive Council estimates that in the United States, reverse logistics costs exceed $35 billion a year.

Over the last couple of decades, the role of logistics has expanded to include recycling, repairs, and returns management. These activities are also known as "reverse flow" or "reverse logistics." The key benefits associated with reverse logistics are to enhance customer service and loyalty and to recover asset values faster. An effective reverse logistics strategy can ensure the recovery of value from returned products. In addition, well-designed logistics strategies can create a significant customer service opportunity and in some cases can also generate revenues and improve customer satisfaction. This is apparent in the automotive industry. The Remanufacturing Industry Council International reports that between 70 and 90 % of the aftermarket goods sold have been remanufactured. This is a market that represents $36 billion a year in sells.

Another aspect of this dimension is the prevention of returns. By creating an effective report system and collecting data companies can analyze the reasons for returns as well as monitor and track trends in the process. When working through a distribution chain, companies can also track vendor performance and, where performance factors are low, can work with those suppliers to improve their quality and resolve issues.

11.3.10 The Cost Attribute

Referring back to Buffet's [3] letter to shareholders where he states that the challenge for managers was to implement strategies than can increase the free cash flow of the business either directly or indirectly, logistics is one of those areas in the business that needs a clear and well-thoughtout strategy because of its significant effect on the supply chain and the operational effectiveness of the company. Transportation costs are frequently a differentiator between suppliers because the higher the transportation costs, the lower the profit margin the company has to work with, and the lower the value proposition they can present to the customer.

There are a number of significant links between logistics and shareholder value. Two of the more obvious links are the impact that logistics can have on both net operating profits and on capital efficiency. The five principle drivers of shareholder value are revenue growth, operational cost reductions, fixed capital efficiency, operating capital efficiency, and tax minimization. Logistics will affect all of these drivers either directly or indirectly.

With revenue growth, logistics can impact both sales volume and customer retention. By insuring that products and services are delivered on-time and to specifications, higher levels of customer satisfaction can be gained, which in turn will drive higher level of sales volume and customer retention. Typically, when customers are happy with the service they receive, they will often repurchase, at larger quantities.

It is generally accepted that cost reduction in operations will result in between a 3 to 1 and a 5 to 1 return (i.e., a dollar saved equals between 3 and 5 dollars growth in net profits). Over the last three decades, it has been recognized that supply chain logistics costs represent a significant proportion of the costs of the final product. Furthermore, there are considerable costs to the firm associated with the internal management and execution of transportation, storage, handling, and order processing.

By its very nature, logistics is a fixed assets intensive activity. This degree of expense to the firm is one reason for the growth in third-party logistics outsourcing. The impact of the dollar intensity on future growth is exponentially higher when one considers the opportunity costs associated with it. For example, if the firm has $10 million dollars invested in warehousing, trucks, and material handling equipment and has a 10 % net profit margin and can flip its inventories 20 times a year, and the value-add of the managing their own logistics versus outsourcing is equivalent to $10 million dollars a year, the opportunity costs (losses) are approximately $10 million dollars a year.

The working capital burdens of managing the logistics systems are high. This problem is compounded by the length of the cash-to-cash recovery time. In many industries, this cycle time is between 6 months and a year. Fortunately, the cash recovery cycle time tends to be directly related to the lead times; thus, reductions in non-value-adding times in the supply chain can dramatically reduce the amount of working capital consumed by the logistics system.

11.3.11 The Process Attribute

Logistics services can significantly influence customer satisfaction and loyalty measures for a company. The dimensions that we presented will have varying effects across the spectrum of market segments. However, all are factors in the customer's quality perceptions of the company and its products and services. As such, the logistics function must be treated as a process, just like production activities. Because of the differing emphasis placed on the dimensions of logistics services by the various market segments, there are many opportunities for companies to customize their systems to increase their competitive advantage.

The process of executing logistical services is a function of procedures and responsiveness. The attribute of procedures is focused upon decision making and methodology. In bounding decision making, management needs to articulate organizational strategic and operational goals and objectives for the various logistics activities. These activities include, but may not be limited to, supply chain coordination and synchronization, risk management, transportation, warehousing, distribution, returns, and production support. In addition to the creation of work methods, material and information flows, and operating procedures, management also needs to establish trade-off priorities. The attribute of responsiveness is concerned with the development and maintenance of flexible capabilities that can address changing customer needs. In actual practice, this attribute overlaps with the flexibility dimension discussed previously.

11.3.11.1 Outsourcing

The outsourcing of logistics functions has proven to be an effective tool in achieving competitive advantage, improving customer service levels, and in reducing overall logistics costs. Previous researchers suggest that firms should take a strategic approach to achieving long-term goals in the outsourcing decision rather than a function-by-function approach. The outsourcing company needs to have a strategic orientation for the partnership. Three other factors influence the service provider's ability to facilitate solutions within the supply chain. These are (1) the perception of the service provider's role within the logistics strategy, (2) the nature of the relationship, and (3) the extent to which the logistics process is outsourced.

Companies following a cost and differentiation strategy often achieve stronger performance than firms using other models. There is usually a substantial cost reduction and customer service improvements that are directly attributable to the 3PL partnership. The study also finds that "highly successful" relationships exhibit several distinctive characteristics: a clear separation of responsibilities, a tiered provider structure, a close working relationship, and a strong performance orientation.

It has long been advocated that in order to maximize their competitiveness in the marketplace, companies should leverage their core competences by forming strategic partnerships in areas of weakness with companies that possess those complementary core competences. It is important that buyers and suppliers match their respective needs with the capacities of the other in order to maximize product differentiation and minimize costs. Many firms have achieved improved quality and timing of purchased products through the reduction of their supply base to just a few highly competent suppliers. Over the past two decades, there have been numerous examples of where suppliers have contributed significantly to the improvement of a firm's products and processes. Research has found a significant positive relationship between JIT purchasing, supplier partnership, and supplier development practices and several performance measurements.

In order to speed up their delivery process and improve customer service, many companies have partnered with transportation specialists to provide and manage activities such as cross-docking and direct store deliveries without the need of inbound inspections. Many companies have adopted postponement strategies to improve their customer satisfaction and competitive flexibility. One of the primary goals of the outbound logistics function in a supply chain is to reduce inventory levels while simultaneously improving customer satisfaction.

11.4 Implications for the Customer

Researchers have found that with respect to logistics service quality, customers are concerned with factors such as person contact quality, order release quantities, information quality, ordering procedures, order accuracy, order condition, order quality, order discrepancies, and timeliness. They also have recommended that firms should customize their logistics programs by customer segments. It has been found that customers valued all of the previously discussed components of logistics quality, but customers in the various segments weigh the value of those components differently.

Studies have also shown that logistics service quality has a positive effect on customer loyalty. In addition, the operational dimensions of logistics service quality (e.g., availability, condition, reliability, and timeliness) and the relational dimensions (e.g., communications and responsiveness) each have a positive effect on customer satisfaction.

Figure 11.2 correlates the visibility and importance of the various quality dimensions for supply chains with the four customer groups. Table 11.1 provides some common metrics for managing supply chains.

Supply Chain Quality Dimensions:		Stakeholders			
		Investors	Employees	Customers / Consumers	Society
Structure	Relationship Management	?	✓	✓	
	Synchronization	?	✓	✓	
	Collaboration	✓	✓	?	
	Technology Management	?	✓	✓	
	Process Management	?	✓	✓	
Product Lifecycle Management		?	✓	✓	✓
Logistics	Timeliness		?	✓	
	Completeness		?	✓	
	Information		?	✓	
	Inventory Availability	✓	✓	✓	
	Flexibility		✓	?	
	Service Experience	?	?	✓	
	Returns	✓	✓	✓	?
	Cost	✓	✓	✓	✓
	Process	?	✓	?	

Fig. 11.2 Supply chain quality dimensions' visibility to the customer

Table 11.1 Key indicators for assessing logistical performance indicators for assessing customer service

Metric	Calculation method
Customer service level	$\dfrac{\text{(Number of SKUs shipped complete by the promised date)}}{\text{(Total number of SKUs ordered)}}$
Back order rate	$\dfrac{\text{(Number of SKUs back orders)}}{\text{(Total number of SKU orders)}}$
Returns rate	$\dfrac{\text{(Number of SKUs returned after shipped)}}{\text{(Total number of SKUs shipped)}}$
Order lead time	$\dfrac{\text{(Time/date order is delivered)}}{\text{(Time/date order placed)}}$
Invoice accuracy	$\dfrac{\text{(Number of order discrepancies)}}{\text{(Total number of shipments)}}$
Indicators for assessing inventory management system	
Lead-time fill rate	Expected number of SKUs short during lead time
Stock out frequency	Number if times a SKU is short within a time period
Unit fill rate	$\dfrac{\text{(Number of SKUs supplied from stock)}}{\text{(Total number of SKU orders)}}$
Inventory accuracy	$\dfrac{\text{(Number of SKUs counted)}}{\text{(Number of SKUs on record)}}$
Percentage of out-of-stock items	$\dfrac{\text{(Number of SKUs out of stock)}}{\text{(Total number of SKUs in inventory records)}}$
Inventory turnover	$\dfrac{\text{(COGS from stock sales during past 12 months)}}{\text{(Average inventory value during past 12 months)}}$
Indicators for assessing order processing	
Costs per order	$\dfrac{\text{(Order processing costs)}}{\text{(Total number of orders processed)}}$
Indicators for assessing warehousing costs	
Average storage costs per location	$\dfrac{\text{(Total warehousing costs)}}{\text{(Total number of storage locations)}}$
Average movement costs	$\dfrac{\text{(Total warehousing costs)}}{\text{(Total number of stock movements)}}$
Indicators for assessing the transportation system	
Transport damage rate	$\dfrac{\text{(Number of SKUs damaged during transport)}}{\text{(Total number of SKUs shipped)}}$
Transport reliability	$\dfrac{\text{(Number of on-time deliveries)}}{\text{(Total number of shipments)}}$
Transport flexibility	$\dfrac{\text{(Number of fulfilled transport specifications)}}{\text{(Total number of shipments)}}$

References

1. Mentzer, J. T., DeWitt, W., Keebler, J. S., Min, S., Nix, N. W., & Smith, C. D., et al. (2001). Defining supply chain management. *Journal of Business Logistics 22*(2), 1–25.
2. Council of Supply Chain Management Professionals. (2010). *Supply chain management: Terms and glossary*. http://cscmp.org/Resources/Terms.asp.
3. Buffett, W. E. (1995). *Chairman of the Board, Letter to the Shareholders of Berkshire Hathaway Inc.* Dated March 7, 1995.
4. Cooper, M. C., Lambert, D. M., & Pagh, J. D. (1997). Supply chain management: more than a new name for logistics. *International Journal of Logistics Management, 8*(1), 2.

Chapter 12
Implementing Change in the Supply Chain

Due to the increasingly high level of competition in our global business environment, most firms are trying to increase their productivity by eliminating problems in their supply chain systems. The objective of supply chain management is to improve the coordination and synchronization of activities between all members in the supply chain. This will help establish a smoother and faster flow of materials and increase the visibility of information along the supply chain. The goal is to improve customer satisfaction through improved flexibility and lower costs. Therefore, structural alignments, policies, and/or practices that impair information visibility or disrupt the flow of materials must be removed from the process.

As a starting point for any improvement activity, one must understand what the business objective is and how the process(es) being improved affect that objective. All too often, people are not clear as to the real objectives or goals of their pursuits. For example, take the question, "what is your company's primary objective with its supply chain?" The answers could range from meeting every order by the requested due date, to being the lowest cost producer in one's chosen marketplace, to maximizing manufacturing throughput, to minimizing everyday working capital, to maximizing market share.

With each of these optimization strategies, only one aspect of the supply chain is being addressed, while ignoring the fact that it is the effectiveness of the entire chain that really matters. Though there are many possible objectives in any business endeavor, the overarching objective for all businesses is long-term profitability. To achieve this objective, it requires optimizing the whole, not just one or more of its parts.

The supply chain, as an extension of the company's operations, is a complex, interdependent system. As a system, decision making must be based on finding a balance between all of the parts and then optimizing that balance. This frequently means suboptimizing one's own function-specific metrics. For example, manufacturing loves long batch runs to maximize the distribution of setup costs

© Springer-Verlag London 2015   161
G.N. Kenyon and K.C. Sen, *The Perception of Quality*,
DOI 10.1007/978-1-4471-6627-6_12

and the lowering of operating expenses. However, doing so reduces flexibility and increases inventories. Both of these results compromise the firm's ability to meeting customer service goals and reduce costs. Conversely, carrying no inventory leaves the production system vulnerable to disruptions and allows sales to take unlimited orders for every possible mix of products. This will slow down a production system with an unmanageable number of changeovers and setups.

The primary purpose of supply chain management is to improve the coordination and synchronization of activities, material flows, and information flows across the supply chain, thus improving the firm's ability to meet customer demands as efficiently as possible. To do this, managers need to have a clear set of objective and a common understanding of how each portion of the supply chain contributes to the whole. In addition, decisions need to be rooted in a structured analysis and improvement process.

12.1 Issues and Opportunities

External customers will rarely if ever see most of the activities associated with managing the supply chain, but the degree of success that a company has in doing so is reflected in both the price of their products and services, as well as in the functional quality of those products and services. These benefits in turn increase the consumer's satisfaction and their perception of the quality of the product or services consumed.

On the other hand, internal customers are both directly and indirectly affected by the effectiveness and efficiency of the company's supply chain management policies, or lack thereof. The principal challenges faced by companies in the management of their supply chains are risk management, effective usage of technology, forecasting, complexity of both the supply chain and of products, supplier and customer relationship management, matching supply and demand, and the design and alignment of the supply chain.

Risk management is a broad topic. It involves being able to define and identify potential risks and develop strategies that either leverage the upsides or mitigate the downsides. One of the benefits of developing a risk management plan at the strategic and operational levels of the company is that those companies that make risk management a priority are less likely to incur major problems related to the scalability and responsiveness to volatile demand. These risk management plans need to address the high probability and critical impact supply chain issues, including supplier quality and performance, price volatility, product and service mix, lack of visibility, distribution infrastructure, transaction costs, and other complex and potentially damaging issues.

As technology becomes more prolific in society, the effective usage of technology will become more critical to the success of the firm. One such usage is related to the leveraging of e-commerce as within the supply or distribution channel. In today's business environment, technology is principally used as a facilitator of transactions by making information more visible, or by providing advanced decision-making support. Over the last decade and a half, a few companies have had success in extending their business model using the Internet.

Amazon started out selling books and other consumer products through an online exchange. Today, they have created a presence as a major seller of industrial products, commanding approximately 2 % of that market channel and growing. Considering what Amazon has accomplished in the consumer products sector, and small and midsize industrial manufacturers and distributors need to take e-commerce seriously. What Amazon has done in the consumer product markets, Dell computers have done in the computer industry. With customers expecting to be severed in more increasingly innovative ways, multichannel networks are becoming a key strategy to containing prices and developing competitive advantage.

Even though it is still expected that the sky is the limit for the potential in technology, it has its limitations in the SCM area. For technology to generate benefits, there must be a set of explicitly stated goals and objectives that are shared by external customers and service providers, as well as internal stakeholders across the supply chain, including warehousing, transportation, and sales. Currently, SCM technologies are the most effective when they are used to support tighter collaboration among the supply chain partners and provide visibility into all key aspects of the business.

As companies expand their offerings in order to remain competitive, while at the same time, meeting all of their other corporate goals and objective, in many instances, more technology, processes, and red tape are added into their operations. All of which increase the cost of doing business. This is even truer for companies with multiple businesses and/or operating units. The problem is that no one seems to have the time to stop and ask: Is this really necessary? Everyone says, "Work smarter, not harder," but management rarely thinks about the layers of unnecessary complexity they are adding to their operations.

In the dynamic, hypercompetitive environment businesses operated today, it is critical that plans are reviewed and updated regularly to insure their relevance and to increase their ability to be successfully executed. Equally important is that management reviews all current and proposed policies to insure that they facilitate the company's ability to be successful and do not act as road blocks to their plans. Furthermore, they need to identify opportunities and methods by which business and operating units can share common processes and technology platforms. Another interesting feedback mechanism is to have suppliers provide report cards on how well the relationship is working. In other words, how easy is it to do business with the company, and where do they see the opportunities to streamline and improve the relationship.

Suppliers are increasingly becoming key extensions to the firm's business model. Not only are they critical enablers in meeting customer demands, but also they are a value knowledge base in the usage of their products and services. Due to this, it is important that management understands the full range of suppliers' capabilities and offerings, as well as their weaknesses. The supplier selection process needs to include a rigorous evaluation of all aspects of a supplier candidate's business, including their financial stability, their organizational culture, and their operational capabilities. Finally, the company needs to routinely audit suppliers on an ongoing basis to insure that they are developing and maintaining the desired capabilities the company will need in the future.

As mentioned previously, many companies suffer from systems that are overloaded with complexity. This causes them to be increasingly ineffective and inefficient. One such area is having too many store-keeping units (SKUs). Virtually every company has products or services that are underperforming. The solution is to establish a disciplined process for regularly auditing and assessing product performance and eliminating the dogs or non-performers, or re-engineering the underperformer with promise.

One of the consequences of too much product complexity is slow-moving and obsolete inventories. Inventory is expensive to carry. Not only are there the traditional carrying costs, such as space rent, insurance, taxes, shrinkage, labor, there are also opportunity costs. Consider your own company's inventories. What is the average level of inventory carried over the year in dollars? What is the average profit margin for products sold? What is the average turnover rate for your inventories? Now compute your opportunity losses associated with tying up that much money over the course of one year;

$$\text{Avg. Dollars of Inventory} * \text{Turnover Rate} * \text{Profit Margin} = \$ \ \$ \ \$\$$$

Thus, another area for improvement is to regularly rationalize inventories against forecasted demands.

One of the main drivers of the mismatch between supply and demand is the metrics that are used. Sales and marketing are typically measured by the revenues they generate, while manufacturing is measured on a cost management basis; these two metrics are frequently in conflict. Sales and Operations Planning (S&OP) is one method that has been successful in matching supply and demand issues. The S&OP process reduces functionally oriented thinking by taking a basic forecast of demands and adjusting it using sale date on events not captured by the data used to generate the forecast, and then adjusting it again using supply chain data on constraints that may prevent the firm from meeting the expected demands: thus, generating a better alignment between expected market demands and planned production.

12.2 Alignment of Product Types with Supply Chain Structure

In 1997, Dr. Marshall Fisher, UPS Professor of Operations and Information Management at the University of Pennsylvania's Wharton School of Business, proposed a framework for matching product types to supply chain design. He speculated that poor supply chain performance was due to a mismatch between the type of products that a company was producing and the structure of its supply chain. He observed that even though technology has enabled several new concepts such as mass customization, lean manufacturing, and agile manufacturing, the performance of many supply chains has never been worse. He further noted that in many cases, costs were increased due to dysfunctional industry practices and adversarial relationships between supply chain partners.

In response to this situation, Fisher proposed a framework to assist managers in designing their respective supply chains. This framework was designed to promote a better understanding of the nature of the firm's products and the type of supply chain structure that would best promote customer satisfaction. The underlying premise of this framework is that when the supply chain is not working properly, it is most likely due to a mismatch between the type product (functional vs. innovative) and the type of supply chain (physically efficient vs. market-responsive) supporting it.

In Fisher's framework, functional products include the commodities that are frequently demanded for everyday life, such as groceries, clothing, and gasoline. These products tend to be stable in their design and functionality over time, with predictable demand patterns, and long life cycles. Because of their stability, competition is frequently strong which leads to low profit margins.

On the other end of the spectrum, there are products that are more mercurial. These products require more innovation and frequent changes to remain relevant in the marketplace. Sometimes, companies can use innovation to achieve higher margins with products that are more traditionally thought of as functional. Innovative products have shorter life cycles and higher levels of unpredictability in the demand patterns. They also have a greater variety of substitute products competing for the same market share. Innovative products require a fundamentally different type of supply chain than functional products.

The attributes that define the two basic types of products are as follows:

- *Product Life Cycle:* Functional products typically have life cycles of two or more years, while innovative products typically have life cycles of less than a year.
- *Contribution to Profit Margins:* Functional products typically contribute less than 20 %, while innovative products typically contribute more than 20 %.
- *Product Variety:* Functional products typically involve 20 or less models, while innovative products typically involve numerous models (more than 20 and often significantly more).
- *Average Forecasting Error:* Functional products typically have forecasting errors of less than 10 %, while innovative products typically have forecasting errors of greater than 40 %.
- *Average Stockout Rates:* Functional products typically have stockout rate of less than 2 %, while innovative products typically have stockout rates of greater than 10 %.
- *Average Forced Markdowns:* Functional products typically have forced markdowns on less than 1 % of items, while innovative products typically have forced markdowns on more than 10 % of items.
- *Make-to-Order Lead Times:* Functional products typically have make-to-order lead times of between 6 months and a year, while innovative products typically have make-to-order lead times of between 1 day and 2 weeks.

There are two distinct functions that supply chain perform: physically and market mediation. The physical functions of the supply chain consist of transforming raw materials into parts, components, and finished goods. It also involves the movement of those parts, components, and finished goods from their points of origin to their next downstream location, and eventually to the consumer. These physical activities incur production costs, logistics costs, and inventory costs.

Market mediation is focused on matching supply with demand. Mediation activities typically include forecasting, new product development and commercialization, and price bundling of products and services. The costs associated with these mediation activities are product markdowns when supply exceeds demand, opportunity losses when demand exceeds supply, and customer dissatisfaction.

The attributes that defined a supply chain and the points of difference between the two basic types of supply chains are as follows:

- *Primary Purpose:* Physically efficient supply chains seek to provide stable, predictable demand at the lowest cost possible, while market-responsive supply chains seek to enable quick response to unpredictable demands with minimal stockout or obsolescence problems, or forced markdowns.
- *Manufacturing Focus:* Physically efficient supply chains seek to maintain high-average equipment utilization rates, while market-responsive supply chains seek to develop buffer capacity to adsorb demand and supply fluctuations.
- *Inventory Strategies:* Physically efficient supply chains seek to generate high inventory turnover rates with minimal inventories, while market-responsive supply chains seek to deploy buffer stocks of parts and finished goods.
- *Lead-time Focus:* Physically efficient supply chains seek to shorten lead times without increasing costs, while market-responsive supply chains seek to aggressively invest in ways to reduce lead times.
- *Supplier Selection:* Physically efficient supply chains seek supplier that are low cost and high quality, while market-responsive supply chains seek suppliers that provide fast and flexibility responsiveness to changing demands with high quality.
- *Product Design Strategy:* Physically efficient supply chains seek to maximize product performance and minimize costs, while market-responsive supply chains seek to use modularity to provide product differentiation.

Given the stability and demand predictability of functional products, market mediation is near perfect in execution; thus, the performance opportunities are in increasing the efficiency of operations and the minimization of costs. On the other hand, the inherent instability and unpredictability of innovative products makes market mediation a top priority. In this environment, choosing suppliers based on their speed and flexibility takes precedence over cost-efficiency.

Though it would seem obvious that innovative products match up best with market-responsive supply chains and functional products match up best with physically efficient supply chains, the difficulty is in the proper identification of product types. This difficulty arises from the fact that products that are physically the same can either be functional or innovative.

When innovation is applied to a functional product, the type of supply chain that would most productively support it will change also. For example, what happened to fast food? In many fast-food restaurants, hamburgers are listed for a dollar on their value menu, along with French fries. But when they are bundled together with a soda, they become a meal and the store often charges more than the combined value meal price for this bundle of products. At Starbucks and other coffee shops, a regular cup of coffee is only a $1.50, but if you add streamed cream and sugar and blend, the price increases to $5.00. By innovating the product or the product/service offering, the value perception of the product/service offering increases. In a 2012 study of 259 US and European companies, the relationship between supply chain fit and financial performance, as predicted by Fisher's framework, was validated. The findings of this study showed that the higher the supply chain fit, the greater the return on assets.

12.3 Alignment of Supply with Demand

Once you have aligned your products with an appropriate supply chain, you will need to rationalize your inventory policies. The first step in this process is to analyze the company's past demand history. Validate, and/or identify, how your customers and product lines differ from each other. Identify and understand how demand varies by season with each market segment, and what drives those changes. The primary focus of this analysis is not the "how much," or the "capabilities to meet demands," but the nature and character of the demand patterns as a whole. Though this exercise should be an ongoing one, most management teams are so busy doing their day-to-day jobs that they do not have the time to manually manipulate the information that is collected routinely by the company to get at the useful information it contains.

Next, create an inventory profile. Over time, organizational inefficiencies can cause enough inventory excesses to pay for the new supply chain planning initiative. By analyzing your inventory and your management practices, you can develop a sense of its relevant attributes such as product, package, warehousing, and manufacturing locations. By analyzing your inventory attributes in relation to shipment quantities and patterns, you should be able to identify stocks that are not required to satisfy demand variability.

With the knowledge gained from your analysis, create a demand planning process that can be systemically maintained and executed. It is highly recommended that the process engages supply and distribution partners in a collaborative manner. With this type of planning process in place, demand dates can be collected effectively and efficiently, thus enabling the company to improve its predictive capabilities in forecasting.

The next step in balancing supply and demand capabilities within the supply chain involves building a quantitative model which reflects your business. This can determine how to support the demand with available assets. This model needs to incorporate constraints and cost factors such as manufacturing facilities, transportation modes, supplier and distribution facilities, and other resources. This model should balance revenues and expenses to find the optimal balance point that is in keeping with business objectives. One of the benefits of such a model is that it facilitates risk evaluation of current and future business opportunities or capacity improvements with greater certainty.

The final step in this process is the development of a systemic sales and operational planning process that will balance supply with demand in accordance with the business strategy given the current constraints on the system. This activity should be implemented as a routine procedure that provides a disciplined approach for responding to changes while minimizing disruption in day-to-day operations.

12.4   The Bullwhip Effect and Inventory Variance

The "Bullwhip Effect" is a supply chain phenomenon that occurs when inventory replenishment practices get "out of synch" with consumer ordering patterns, resulting in supply distortions that propagate upstream in an amplified form. In layman terms, this effect occurs when purchase order volumes to suppliers tend to have greater variance then sales to customers, and these distortions are propagated back up the supply chain. These distortions are exacerbated by many of the common practices used to adjust inventories such as backordering, promotions, etc. When this situation occurs, there are direct, and often significant, impacts to production schedules, inventory control, and delivery plans between various supply chain members, and ultimately the consumer. In 1993, Kurt Salmon Associates, a management consulting firm serving the retail, consumer products, and healthcare industries, estimated that these types of problems could result in additional operating costs of between 12.5 and 25 %.

12.4.1   Primary Causes of the Bullwhip Effect

Because consumer demand patterns are rarely stable, businesses must forecast their expected demands in order to properly position their inventories and resources. The art of forecasting involves the use of qualitative (e.g., consumer surveys, expert opinions, and Delphi studies) and quantitative methods (e.g., statistics, simulation) to model and predict what is most likely to happen in the future. These methods are not perfect. Thus, forecasting errors occur. These errors drive companies to create inventory buffers, called "Safety Stock."

As the end consumer's purchasing history is disaggregated into its components as the information is past back up the supply chain, forecasting errors are compounded by additional forecasting errors. This compounding of errors in turn induces suppliers along the way to create additional inventory buffers. When these buffers grow to excessive levels, companies will take actions to reduce them, which in turn will add addition volatility to the demand signal going to suppliers.

The behavioral causes of a bullwhip effect are generally driven by management behavior in the downstream companies (e.g., those closes to the end consumer) of the supply chain. Though the human factors influencing these behaviors are largely unexplored, studies suggest that insecure people (e.g., needing safety and security) seem to perform worse than risk-takers. Additionally, people with high self-efficacy typically have less trouble handling chaos and variability. Some of the primary behavioral causes are as follows:

- Improper usage of base-stock policies.
- Misunderstanding of feedback data and time delays.
- Over-reacting to demand variances, such as panic-driven ordering after unmet demand.
- Projecting opportunistic intent into the actions of others, such as perceived risk of other players' bounded rationality.

Operational causes of a bullwhip effect are associated with the actual practices used in executing normal business activities. The primary practices used to mediate and reduce the bullwhip effect are order batching, promotions, price variances, lead times, and rationing and gaming the system. In isolation, these practices are often beneficial to the company they occur in, but in the dynamic environment of a supply chain, they can result in costly and at times catastrophic problems.

Order batching refers to the practice of placing orders into the supply chain in batches in order to leverage economics of scale. The variance in normal ordering patterns caused by this action will send shock waves up the supply chain, and when this behavior persists, suppliers will interpret the actions as a permanent change shift in demand. As the batching increases, so does the shock waves across the supply chain.

Price variances and promotions refer to the practice of incentivizing customers to increase their purchases by reducing prices. However, this often results in shifting the timing of their purchases. This shifting of demand, if continued over time, frequently results in spiked demand patterns that are not in alignment with actual consumer demands.

Lead times are determined as the lapsed time between when an order is placed and the associate purchases and or deliveries are received.

- *Rationing and Gaming* occurs when supply shortages cause customers to over order, or to increase safety stocks, to counter the effects of the shortages. This type of action will in turn increase the load on upstream resources exacerbating the supply constrained situation and cause customers to increase their avoidance behaviors.
- *Demand Signal Processing* refers to the updating of forecast data with past demand information. The issue with this practice is the assumption that what happened in the past will happen in the future. The problem occurs when past actions need systematic analysis keeping in mind actual customer behavior, such as reactions to product promotions, or limited time price discounts, or inventory reductions. When these types of data exist in past sales data and are not properly formatted to remove their affects in forecasting, they will skew the

results causing additional forecasting error. Other actions that can skew forecasting results are supply constraints that limit production and sales opportunities. Similar results can occur when planners perceive a future event as driving increased sales, and then, the sales do not meet those expectations. Thus, future orders are reduced for a period to exhaust the excess inventories.

12.4.2 Consequences of the Bullwhip Effect

There are several consequences associated with the Bullwhip Effect: greater safety stocks, inefficient production, or excessive inventory. Furthermore, the constant hiring and dismissal of employees to manage the demand variability induces additional costs due to increased training, possible lay-offs, and lower morale which will often leads to lower productivity. All of these problems will lead to lower utilization of distribution channels, potential stockouts, and poor customer service.

12.4.2.1 Excessive Inventory

Increased volatility of demand causes greater inaccuracies in forecasts. This in turn causes increased inventories and stockout alternately. Demand volatility also causes inefficiencies in production systems. Increased inventories and lower operational effectiveness not only drive increased costs, but they also promote the offloading of excess inventories in order to generate more revenues. These products are most likely to be deeply discounted which not only reduces margins; it also slows down customer reorder rates which again result in increased demand volatility. Reductions in order lead times and increased visibility of true customer demand patterns can increase forecast accuracies and reduce or eliminate these types of problems.

12.4.2.2 Inefficient Production

The demand volatility associated with the Bullwhip effect creates production inefficiencies by reducing flexibility and disrupting planning. Good planning practices require relatively stable demand patterns. When demand unexpectedly spikes up the production system, the manufacturer cannot produce enough products ahead of time and cannot schedule production in an efficient way. The potential of stockouts increases dramatically. When stockouts do occur, there are lost revenues from the missed sales. Furthermore, there is a loss of reputation that can result in reduced future demands.

This causes management to shift into a reactive decision-making mode, and they will rush to produce more by increasing production rates about the optimal operating level, working overtime, or outsourcing more work: all of which will drive costs up. The increase in work load will drive the need for hiring and training of extra staff, increase wages from overtime, and increase procurements costs due to the push to get materials quickly. Then, to compound the problem, when demand decreases, they will build higher levels of safety stock, hoping to avoid the problem again in the future: again, increasing costs.

12.4.2.3 Increases of Cost

As we have pointed out, the most damaging effect of the Bullwhip Effect is increased costs. Given that most companies have limited opportunities to increase product prices because of competition, as their operating expenses increase, their profit margins will decrease. With decreased profits, there is less money with which to maintain and grow the business, develop and introduce new products, improve quality and any number of other things that make a company more competitive.

12.5 Counter Measures for the Bullwhip Effect

Traditionally, sales data and inventory status data have been considered proprietary to the individual companies. As such, there was no obligation to share those data with others. One prescription for overcoming the bullwhip effect is that these data need to be shared with the manufacturer. By making customer demand (e.g., point-of-sale data) data available across the distribution side of the supply chain, significant improvements to forecasting can be made. In other words, increase visibility along the supply chain. Even then, there may still be problems due to differences in forecasting methods and buying practices. Research has shown that when downstream supply chain partners make their inventory and demand information available to upstream partners, significant improvements can be made in inventory control across the chain.

Even with improved visibility along the supply chain, another impediment to smoothing the material flows is the batching of orders by downstream members to counter the costs of ordering, replenishment, and less-than-truckload transportation. If the manufacturer offers multiple products, one countermeasure is the order a variety of products in batches much closer to actual demand levels, so as to fill the truck. The use of third-party logistics firms capable of economically dealing with less-than-truckload orders is another alternative.

A second impediment to smoothing the material flows in the supply chain is variable pricing. One approach to helping retailers and suppliers eliminate the promotion and discounting practices that cause demand variability is for manufacturers to establish uniform wholesale pricing, thus reducing the incentives toward forward buying.

There are situations where the problem is not with the retailers or wholesalers, but with the dysfunctional decision making at the manufacturer. Several years back, I worked for a major computer manufacturer in Houston, Texas. This company's dysfunctional marketing practices had so distorted their demand patterns that forecasting accuracy at the product family level was in the low 70 % range. The driving issue behind the problem was upper management trying to manage their quarterly stock price. Whenever it appeared that quarterly sales were going to be less than forecasted, they would push marketing to incentivize their channel partners to buy more by offering price discounts or quantity discounts. After several quarters of incentives, the company had successfully trained their distribution partners into waiting till the end of the quarter to place their orders in hope of receiving a price discount. This buying practice by the customer virtually insured that quarterly sales would be low enough to cause a stock downgrade unless management discounted the products to capture more sales revenues.

12.6 Implications

Brand recognition, customer loyalty, and company reputation are valuable commodities that are hard to earn and easily lost. Being able to product and deliver quality products that customers value, quickly and at a fair price is the best way to earn brand recognition and customer loyalty. But to do that, the company must develop a reputation of being a well-organized, technically competent, quality-focused organization that is fair and flexible to work with in order to hire the best people and attract the best supply partners.

Developing a world-class enterprise is not easy; it requires a properly designed and managed supply chain. This only happens by selecting the right suppliers and continuously driving value-adding improvements as we discussed. Being fair, being trustworthy, and being smart are mandatory. Any other approach will create negative perceptions in both our internal customer base and external customer base.

Chapter 13
The Dimensions of Product Quality

Quality is first and foremost a strategic question, in that it governs the development of product designs and the choice of features or options, as well as setting the criteria for the selection of suppliers and materials. Because product quality is a major factor in the development of a sustainable competitive advantage, most companies address objectives for improving product, process, and service quality at the strategic level as a method for achieving world-class performance. In fact, customer impressions about a firm's products are frequently formed from past experiences with the firm and its products.

With respect to quality management and the design of products or services, managers need to think strategically about quality and focus on those dimensions of quality that support their strategic objectives. There are several dimensions associated with product quality: performance, features, reliability, conformance, durability, maintainability, aesthetics, and innovativeness. These dimensions are often interrelated such that improvements in one dimension might be at the expense of another.

13.1 The Performance Dimension

From the customer's perspective, the choice of purchasing a product starts with the desire to satisfy a given need. Thus, the product must possess specific performance characteristics or capabilities. The term "performance" refers to the primary operating characteristics established by the design of the product. For example, the primary characteristics of a car are acceleration, handling, cruising speed, comfort, convenience of features, and braking distance. The degree to which individual customers will classify these characteristics as either good or bad quality will depend upon where they are in a wide range of the interests and needs.

© Springer-Verlag London 2015
G.N. Kenyon and K.C. Sen, *The Perception of Quality*,
DOI 10.1007/978-1-4471-6627-6_13

13.2 The Features Dimension

Coupled with the product's performance are the features by which this performance is achieved: the "bells and whistles." If performance is considered the primary set of characteristics of a product, then features must be considered the secondary set of characteristics. Like with performance characteristics, features possess objective and measurable attributes. With a car, these features may include setting for four, a digital radio, surround sound, GPS, dual air control, power steering, power windows, power seats, etc.

Usually, more features are perceived as equating to greater performance, but there are exceptions to this generalization. For example, fighter aircrafts that are designed to perform multiple missions (e.g., dog fighting, bombing, and close air support of ground troops) are frequently less to superior at any of these missions because of the numerous trade-offs in performance needed to accommodate all of the desired functions.

13.3 The Reliability Dimension

Reliability is defined as the probability of a product performing within specification over a specified period of time. Common measures indicating the reliability of a product are the mean time to first failure (MTFF), the mean time between failures (MTBF), and the failure rate per unit time. The data collection process for these two measures require that the product be in operation for a period of time; either in a test environment or in actual field usage. As a result, most reliability measures tend to be most applicable to durable goods than to products and services that are consumed.

In the design of products, there are basic systems configurations for which reliability can be mathematically calculated: series, parallel, mixed, standby redundant, and k out of n. Gain the information feed for these calculations require that data be collected on the product and/or components while they are in operation. The probability models for these basic system configurations are as follows:

$$\text{Series Configuration: Rsys} = \prod_{i=1}^{n} P(X_i)$$

$$\text{Parallel Configuration: Rsys} = 1 - \prod_{i=1}^{n} (1 - P(X_i))$$

$$\text{Mixed Configuration: Rsys} = P\big(B_1 \cup B_2 \cup \cdots \cup B_j\big)$$

where $P(X_i) =$ the probability of component success, $(1 - P(X_i)) =$ the probability of component failure, and $B_j =$ the configuration of the jth branch.

13.4 The Conformance Dimension

Conformance is often measured by the degree to which the elements of a product's design and operating characteristics meet the parameters of a pre-established set of standards. From a production perspective, conformance can be measures by various metrics such as yield rate. From the field service perspective, conformance is often measured using proxy metrics such as the incidence of service calls and the frequency of repairs or replacements under warranty. From the customer's perspective, conformance is measured by the degree to which the product's design and operating characteristics meet their expectations for the product.

Customers will usually form a set of expectations about these quality dimensions that best meet the circumstances of their needs, their experience in solving similar needs, and information received directly and/or indirectly. Thus, the degree of conformance between the products actual performance against these expectations, or standards, is important to their perception of the product's overall quality.

13.5 The Durability Dimension

Durability is an operational dimension that is closely related to reliability and to conformance. Durability is a measure of a products useful life and is commonly measured along two aspects: technical and economical. From the technical aspect, durability defines the amount of usage one receives from the product before its physical deterioration. From the economic aspect, durability is a perceptual measure of value based upon the amount of usage received, before the marginal value of the next usage is sufficiently less than the replacement costs of the product. Often both the technical and the economic aspects can be extended when the product is repairable.

From the previous discussion, it can be seen that durability is closely linked with reliability and conformance. If the product fails to conform, or is within marginal conformance, it will be unreliable and will prematurely fail. Products that fail, or need to be repaired frequently, are very likely to be replaced more quickly than those that are more reliable.

Durability metric must be analyzed carefully before being using in decisions. One reason is that products may be in service beyond their expected useful life due to economic reasons instead of technical improvements. For example, during economic recessions, money is often tight and economic security is frequently low; thus, people tend to hold onto their durable goods longer. Where we might replace a 4 or 5 years old car in good times, we might keep that car an additional 4 or 5 years till the economy improves. On the other hand, if our car was a large gasoline guzzling tank, when fuel prices significantly increase and there is the expectation that they will stay high, the tendency is to replace the gas guzzler with a fuel efficient model sooner than planned.

13.6 The Serviceability Dimension

Over time the maintenance and serviceability of a product will moderate the customer's perceptions of the product quality. Those products that are difficult or expensive to maintain or service, the perception of quality will frequently be downgraded, while those that are not, will be upgraded. One way of addressing this concern, especially with commercial customers, is with post-purchase support for the product.

Garvin in his ground breaking publication of "What does 'Product Quality' really mean?" describes serviceability as the speed, courtesy, and competence of repair. This description implies that others are providing the maintenance and servicing of your products. Personally, it irritates me when a product that I purchase is so complex that I am unable to perform even the basic servicing necessary to keep it in working order, much less perform actual maintenance on it if it needs repairs. It was not that long ago that an individual was able to perform the basic servicing and maintenance needed to keep their own cars working properly. Today, our cars are much more expensive and so complex that it is virtually impossible for the individual to maintain their own car anymore, even including moderately competent individuals who have worked on previous models. Today, even the simplest of repairs require you to take your vehicle to a certified mechanic and spend hundreds of dollars. I don't consider that to be "Good Quality."

Service is defined as someone else doing work for you. As such it is not a characteristic, much less a dimension, of a product's design. Thus, this product dimension must be re-defined as follows: serviceability is measured by the degree of speed and ease with which the product can be serviced and maintained.

There are two attributes of this dimension are as follows: (1) the ability to service the product and (2) the ability to maintain the product. The servicing of a product involves the cleaning and replenishment of any items consumed during the operation and usage of the product. For example, cars periodically need to replenish or change the engine oil, transmission oil, power steering, brake, and cooling fluids. Servicing also includes the periodic adjustment of feature setting, such as brakes or the engine idle, so that the product will continue to perform at an optimal level. Maintenance involves the repair and replacement of broken and/or damaged parts or components that are elements of the product's design. Without these repairs, the product would either be unsafe to operate or would fail to operate at all.

13.7 The Aesthetics Dimension

The measurement of a product's aesthetics is purely subjective. It is a personal judgment, based upon individual preferences of whether or not the product looks, feels, smells, tastes, or sounds appropriately. If a product is designed correctly, it should stimulate emotional, hedonic, and practical benefits for the customer.

Research into product design methods has found that the aesthetics of a product can create feelings of either happiness or anger, pride or shame, security or apprehensive. When the design creates positive emotional responses, it adds additional value to the product. Ideally, design should create an emotional linkage between ideas, products, and brands.

Karl Duncker, an early twentieth-century Gestalt psychologist, identified three types of pleasure: sensory, aesthetic, and accomplishment. He defined sensory pleasure as being derived from satisfying a craving for stimulation of the optical, olfactory, and acoustical receptors. The sensation can also be unpleasant when they stimulation frustrates some other conation such as a balance of function. Aesthetics pleasure on the other hand is a sensory pleasure that is derived from a perceived meaning that is being expressed by an object. "Thus the appearance of a weather-beaten tree may express undaunted tenacity; the stroking of a hand may express tender solicitude; a Mozart Rondo may express sprightliness and gaiety; a wide view may express infinity, and widen our hearts. Aesthetic enjoyment is the principal, though not the only instance of enjoying something expressed in the process of expression" [1]. Of the three types of pleasure, only sensory and aesthetic pleasures can be derived from visual appearance of a product.

13.8 The Creativity Dimension

Creativity has been described as the synthesis of new ideas and concepts, whereas innovation is the implementation of creative ideas. For a company to be successful with the introduction of innovative products, they must have a clear and significant

Product Quality Dimensions:	Stakeholders			
	Investors	Employees	Customers / Consumers	Society
Performance	✓	?	✓	✓
Features		✓	✓	
Reliability	✓	?	✓	✓
Conformance	?	✓	✓	
Durability	✓	?	✓	✓
Maintainability		?	✓	✓
Serviceability		?	✓	✓
Aesthetics	✓	✓	✓	
Creativity	✓	✓	✓	

Fig. 13.1 Product quality dimensions visibility to the customer

point of difference from their competitor's offerings. The Austrian economist Joseph Schumpeter defined five aspects associated with innovation [2]:

1. The introduction of a product, which provides consumers with a novel approach to solving their needs, or one of significantly higher quality than other products that have been made available in the past.
2. The development of new methods of production, relative to a particular industry segment.
3. The development of new sources of supply.
4. The development of new markets.
5. Novel new approaches for marketing products and service that leads to the restructuring of an industry.

Of these aspects, the first three that are the most relevant to the design of the product, while the last two are the most relevant to the commercialization of that product.

Figure 13.1 provides a mapping of the quality dimensions that define our perceptions of a product to the customer/stakeholders that are vested in the product.

References

1. Duncker, K. (1941). On pleasure, emotion, and striving. *Philosophical and Phenomenological Research, 1*, 415.
2. Schumpeter, J. A. (1943). *Capitalism, Socialism, and Democracy* (6th ed., pp. 81–84). New York, NY: Routledge.

Chapter 14
Product Design and Commercialization

There are three characteristics to every design: visceral, behavioral, and reflective. These three characteristics are interwoven through every design, eliciting both an emotional and a cognitive response. Our emotions are inseparable from and are a necessary part of our perceptions (i.e., cognitive functions). Emotions are also a necessary part of life because they affect how our feelings, behavior, and thoughts. In fact, without an emotional response to different situations, our decision-making abilities would be seriously impaired. Our emotional response to our immediate environment is how information is transmitted to our conscious awareness. For example, fear is an emotional response to a dangerous situation, and joy is the response to pleasant situations. Some objects, or situations, we encounter will evoke a strong emotion such as love or hate, happiness or disgust. Take for example how you felt the first time at seeing an original Shelby Cobra GT, or a photograph of the Grand Canyon at sunrise. Now compare that to how you would feel while touring the Holocaust Museum, the Viet Nam War Memorial.

When designing a product, characteristics such as the aesthetics, attractiveness, and beauty must be considered. Many designers get caught up in the functionality of the design and overlook its beauty and aesthetics. Beauty and functionality are not diametrically opposing attributes. In fact, pleasure and usability go hand in hand.

For several decades after its invention, televisions had only black-and-white screens. Even though we see the world in color, watching television in black-and-white was still immensely enjoyable. When color screens where developed, television sets became heavier and more expensive, but the enjoyment factors associated with viewing television shows increased considerably. Today, if you gave someone a black-and-white television, they would probably think that you were insulting them. From a cognitive perspective, the color does not add any real value; from an emotional perspective, there is a significant improvement in the viewability of the shows.

© Springer-Verlag London 2015
G.N. Kenyon and K.C. Sen, *The Perception of Quality*,
DOI 10.1007/978-1-4471-6627-6_14

The main problem that designers face when creating a new product is that they often make decisions based solely upon logic, even if their emotions are telling them otherwise. Our emotions pass judgment on things that we experience, thus providing us with information about our environment and facilitating our ability to make decisions.

Affective and cognitive systems are our mind's information processing systems. Our affective system makes quick judgments about the good and bad elements in our environment, while the cognitive system interprets these signals and makes sense of them for our world perceptive. Furthermore, affect and emotion are critical to decision making. Research has shown that our affective system enables us to quickly choose between good and bad alternatives. However, without emotions, we are often unable to actually make a final decision.

Emotions are our conscious experiences of the affective signals and complete with attribution of its cause and identification of its source. As such, our emotions are tightly coupled with our behavioral response mechanisms, preparing us for a physical response to a given situation (i.e., a flight or fight response to bad situations or embrace in good situations). In order to create a successful product, the design must not only be functional, it must also elicit an emotional response.

An interesting phenomenon of design is that attractive products are perceived as being easier to use. Understandably, aesthetic preferences of the user are culturally dependent. The basics for this perception revolve around our emotional system changing how our cognitive system operates. Given this physiological response, it is easy to understand how the aesthetics of a design can be perceived as being easier to use.

Research has also found that positive emotions are critical to learning, curiosity, and creative thought. Being relaxed and happy broadens our thought processes, thus expanding our thought processes and allowing us to be more creative and imaginative. Conversely, when we are nervous, or anxious, our thought processes tend to narrow. This leads us to focus more closely upon what is directly relevant to the problem. The conclusion to be drawn here is that attractive items make people feel good, causing them to think more creatively, thus making it easier for them to find solutions to the problems that they are encountering.

14.1 The Perception of Product Quality

There are three distinct perceptual properties that are present in every product offering: search, experience, and credence. The search properties include factors that are related to those characteristics that can be determined prior to actually using the product. The experience properties include factors that are related to those characteristics that can only be evaluated after actually using the product, while credence properties on the other hand incorporate those factors that the consumer often find difficult to judge even after using the product.

The factors that we perceive and use in assessing the quality and value of a product are called dimensions. These dimensions span all three of the perception properties. The strength, or influence, that a given dimension will have in association with a perception property will vary from person to person, and from product to product.

Assuming that there is a baseline level of influence that each dimension exerts in each property, we conducted a survey asking participants to rate the degree of importance that they attached to various quality dimensions for a product with respect to the three properties of perception creation: credence, search, and experience. Theorizing that people would merge these quality dimensions into a unidimensional framework, the findings were then grouped into categories using a factor analysis procedure. The results of this analysis are shown in Tables 14.1, 14.2, and 14.3.

In Table 14.1, it can be seen that consumers will tend to think about product's quality dimensions along two specific paths: the acceptability of the product to perform the desired tasks and the degree of satisfaction to be received from the product's performance. The greater the degree to which a product's ability is believed to deliver benefits that accentuate the quality dimensions of aesthetics,

Tables 14.1 Credence categories of product quality dimensions

Variables	Factor 1 acceptability	Factor 2 satisfaction
Aesthetics	−0.00051	0.75667
Conformance	0.21366	0.64414
Convenience	0.29667	0.62105
Creativity	0.4314	0.46433
Durability	0.46778	0.58291
Ergonomics	0.45192	0.48838
Features	0.46163	0.4933
Performance	0.65499	0.33754
Reliability	0.77383	0.24205
Security	0.80649	0.14691
Serviceability	0.76503	0.16966

Tables 14.2 Search categories of product quality dimensions

Variables	Factor 1
Aesthetics	0.61981
Conformance	0.68223
Convenience	0.69578
Creativity	0.59959
Durability	0.71372
Ergonomics	0.69555
Features	0.59811
Performance	0.70173
Reliability	0.71392
Serviceability	0.66380

Tables 14.3 Experience categories of product quality dimensions

Variables	Factor 1 design	Factor 2 enjoyment
Aesthetics	0.65822	0.23917
Conformance	0.61305	0.15092
Convenience	0.63836	0.31724
Durability	0.81958	0.13027
Ergonomics	0.48636	0.40338
Performance	0.70907	0.26329
Reliability	0.62723	0.35798
Creativity	0.21982	0.68950
Features	0.44869	0.49913
Security	0.24918	0.74337
Serviceability	0.46459	0.47114

conformance, convenience, creativity, durability, ergonomics and features, the more likely that the customer will depend on the product for his usage. The dimensions of performance, reliability, security, and serviceability create the customer's perceptions of satisfaction with the product. In other words, the quality dimensions of aesthetics, conformance, convenience, creativity, durability, ergonomics and features act as "Order Qualifiers," while the dimensions of performance, reliability, security, and serviceability act as "Order Winners."

Table 14.2 shows consumer's thoughts about product quality dimensions alone a single path. Depending upon the critical-to-quality (CTQ) features of the product, developing information on the product and its various dimension of quality is important. This is typically the last opportunity the company has to inform the customer of the benefits of the product.

Just as with the credence properties, while experiencing the use of a product, customers will typically think about its quality dimension along two paths: how well does the design fit the application, and how easy is it to use. The design factors (e.g., aesthetics, conformance, convenience, durability, ergonomics, performance, and reliability) tend to act as "Order Qualifiers," while the enjoyment factors provide the "WOW!"

The above results suggest that manufacturers must emphasize the aesthetics dimensions of their product in communications through their promotional material. These factors captured in Factor 2 are critical in achieving the trial rate for usage by new customers. Advertising complemented by sales promotion tools to lower the risk of consumers trying out new products will be important parts of these promotional campaigns. However, the emphasis on the functionality of the product in terms of features and in terms of their performance, consistency and reliability are essential to finally sealing the deal. Without these qualities, consumers will feel let down and a high trial rate might be followed by a low adoption rate. Thus, the types of qualities, epitomized by the two factors are both essential in providing customer value. Attentions to creating promotional campaigns that emphasize the aesthetic value of a product followed by intelligent product design and a manufacturing process that delivers on the promise through actual performance are therefore fundamental to overall product strategy.

The more robustly a product delivers benefits sought by the customer, the greater the customer's perceptions of the product's quality. The more information the customer receives about the performance and benefits of competitive substitute products, the more likely that the formation of both their perceptions and expectations or to be influenced. Environmental conditions can affect the way customers will perceive the benefits received from their purchases, as well as influence their expectations about the possible benefits of future purchases. By monitoring the customer's reactions to the various attributes of similar products, coupled with an understanding of how those reactions relate to the design of product features, the firm should be able to continuously produce superior offerings for the marketplace.

Each of the characteristics encapsulates a given dimension of quality that has credence, search, and experience properties. These characters align with these perceptions properties at different levels. The model in Fig. 14.1 shows how consumers create their perceptions and how the product characteristics will align with the properties of perception.

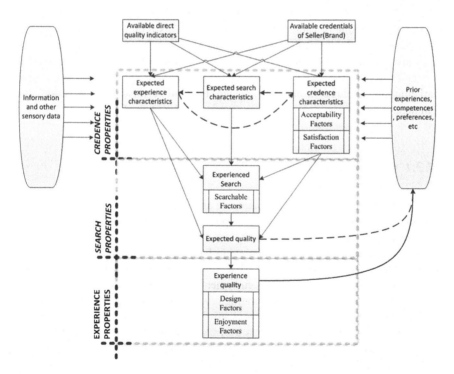

Fig. 14.1 The quality perception process model for products

Another key issue with the design of products is sustainability. During the design process, social, economic, and environmental concerns need to be addressed simultaneously and holistically. Though sustainability is primarily a process issue, the design of the product can significantly affect manufacturing's ability to meet the objectives of a sustainable strategy. To support environmentally focused sustainability strategies, designs must allow for environmentally friendly materials, processes, and practices; including the return and disposal processes.

14.2 The Product Development Process

Successful products do not often just happen; they usually evolve over time through a complex product development process involving strategy, management, research and development, production, and marketing. It also requires the linking of science, technology, and innovation to the marketplace. The process is a goal-directed problem-solving activity which relies heavily on human experience, creativity, and knowledge. To achieve the maximum benefits from this activity, the process needs to integrate creativity and innovation tools into the process.

The objective of the product development process is to design and produce a product that customers value and will purchase. This requires that the work not only consists of logical rules bounding the design procedures, but also needs innovative thinking in order to produce creative solutions. The process itself is a complex set of integrated activities that typically consists of six major stages: marketing and evaluation, planning, design, test and verification, production, and commercialization.

14.2.1 Generating Ideas

It has been found that for every sustainable and commercially viable product, two thousand raw ideas are required. Where do these ideas come from? Basically, ideas can come from anywhere. When there are changes in society brought about by factors such as new laws, companies going bankrupt, new technologies, changing demographics, etc., a void is created. This creates an opportunity for a new product. Many new product ideas come from customers, suppliers, and employees, but most originate from a structured idea generating process in marketing and engineering.

The idea generation process is a creative activity that involves the imaging of something new. Creativity itself is a synthesis of new ideas and concepts that evolve from the restructuring and re-associating of existing ideas and concepts.

Creativity in the design process requires creative thinking balanced with practical considerations. There are numerous creativity stimulation techniques. The following list names a few:

- Bug listing;
- Goal versus wish listing;
- Manipulative verbs;
- Nominal group technique;
- Brainstorming;
- Visual confrontation;
- Morphological techniques;
- Idea diagram;
- Reverse perspective takes a look at traditional assumption and then look at them again from a diametrically opposite perspective.
- Merge multiple perspectives:
- The fresh perspective approach involves describe the problem that is being solved to someone else who is not involved in the design process. Often, they are able to provide new insights and a new perspective.
- Mind mapping is a technique involving the drawing of a diagram for the organization of your thinking. This diagram facilitates the envisioning of how various parts are connected.
- BruteThink is a technique based on random stimuli and involves bringing a random word into the problem (from a dictionary, newspaper, book…) and then thinking of things that are associated with that word. One can even force connections between the random words and then try and make sense of the combined words.
- Bionics involves the search for an existing solution within nature that can be adapted to solve the problem under consideration.
- Checklisting involves the use of words and questions to trigger creative thoughts. These triggers often focus on possible changes to an existing product, concept, or system.
- The use of analogies.
- Adaptation is an activity which involves looking for ways that existing designs can be used to generate new solutions to unrelated problems or that old (and rejected) concepts can be revamped into useful forms.
- The use of routine mental exercises can keep an individual's creative powers in top form: solving word games, scrabbles, anagrams, crosswords, solving puzzles, playing chess and checkers, etc.
- Using analogies and metaphors to trigger ideas, making the familiar strange and the strange familiar.
- Inversion is to concentrate on ways to make a product or system less effective and then invert these ideas to form ways in which the product can be improved.

Some researchers of creativity consider the creative process as a five-step one: fact-finding, problem-finding, idea-finding, solution-finding, and acceptance-finding.

Others suggest a simpler way, recommending twenty ways in which you can stimulate your own creativity:

1. Keep a notebook and pen with you at your desk, in your car, everywhere you go so you can always record your ideas as you have them.
2. Learn to ask more questions. Ask "What if?" Don't be afraid to have what seem to be completely crazy ideas. Learn to think beyond what seems obvious.
3. Daydream. Learn to use your imagination. You must see it in your mind before it becomes a reality. Become a master at visualization.
4. Become more spontaneous. Don't concern yourself with what others think about your ideas.
5. Follow your intuition. Follow that little voice inside; it will never let you down.
6. Do not compare yourself to others. This is a quick creativity killer. Be nice to yourself.
7. Try new things. Visit a museum, go shopping, a change of scenery will reboot your creative mind.
8. Brainstorm with your family or friends.
9. Read, read, and then read some more. Whether you are reading a book, your tweets, the Internet, just read. Never stop learning.
10. Engage in a non-stress, relaxing activity that does not require thinking, such as watering the lawn and taking a bubble bath.
11. Learn to become comfortable with the silence. Quiet your mind. Go within, there are nuggets of gold in there if you can quiet your mind and explore yourself. Get in the moment.
12. Try new foods. Milk, spinach, salmon, whole-grain pasta, tofu, and sunflower seeds are all proven to make you feel happier and more creative.
13. Sleep or lack of it. It is funny because it is shown that sleep can recharge your batteries and make you feel more creative, but also hacking your body and not sleeping for extended periods of time can trigger insane amounts of creativity.
14. Listen to music. (This one is my favorite!)
15. Physical exercise definitely stimulates creativity.
16. Take a walk in nature, breathe in the fresh air, appreciate, and become inspired by nature.
17. Stop doubting your own ability to be creative. Believe that you are inherently creative.
18. If you hit a wall creatively, walk away and do something else for a while, then go back to your project. When I do this, the answer almost always magically comes to me.
19. Learn to take more chances, put yourself out there a bit more, trusting yourself.
20. Last but certainly not least, fall in love with what you do. All creative people have this in common. For example, I love to write. I write for the same reason that I breathe, because I have to. It is who I am and I love it. Find your passion and do it. And then, do it some more.

Creativity is a complex interdisciplinary construct influenced by a 'combination of personality (people), outputs (products), sequence of tasks (process), and environmental setting (press). It is also the hallmark of highly successful companies.

14.2.2 The Market Evaluation Stage

As the firm's face to the world, marketing is responsible for all direct interactions with the marketplace. With respect to the product development process, marketing has two important functions: marketing and evaluation and product commercialization. The marketing and evaluation stage of product development typically involve the idea generation, a market feasibility study (e.g., a strengths, weaknesses, opportunities, and threat analysis), problem definition, and information gathering. The commercialization stage will be discussed later.

Total quality management philosophy has long espoused the capturing and incorporation of customer requirements within product design. The process of capturing of these requirements involves understanding decision regarding not only of what type and how much information to gather, but also firm-level decisions on which market segments to pursue, competitive strategies for those segments, and resource availability.

The impetus for a new product comes from perceived opportunities in the marketplace. These opportunities will be driven by either a need currently fund in the marketplace (e.g., market pull), or a potential need in the marketplace due to the advent of a new technology (e.g., technology push). The development of the product concept is crucial to the success of the product over it projected lifecycle. The steps associated with concept development are identifying customers and the requirements, establishing target specifications, identifying competitive products, generating alternative product concepts and selecting the best option, refining the product specifications, performing an economic analysis, and then formalizing the development plans for the commercialization of the product (Fig. 14.2).

14.2.2.1 Identifying Customer Needs

Methods for the collection of customer and consumer data include interviews, questionnaires, feedback, databases, and other methods. It is often difficult to collect accurate data about what the customer base wants and expects, generally because customers and consumers frequently are unsure about what it is they really want. They typically understand what benefits they want from a product, but often have only a vague idea about how to get that benefit. In many cases, customers may not even be aware of solution they would value because they have dealt with a set of problems for so long they just accept them without question. As a result of this lack of understanding, multiple probing inquiries are necessary.

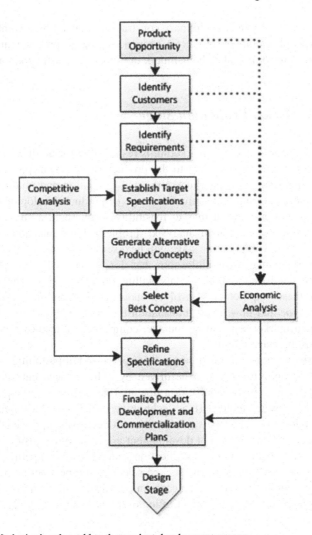

Fig. 14.2 Marketing's value add to the product development process

Another consideration during the data collection process is determining how various customers perceive different functionality, features, and quality characteristics.

14.2.2.2 Establishing Target Specifications

Once customer requirements have been captured and analyzed, the information needs to be translated into a set of functional specifications that engineering design can use for the detailed design. Though this activity is basically a marketing function, product designers and engineers often become involved in the process.

Quality functional deployment (QFD) is a well-documented technique for this process. When factoring in the additional knowledge of how customers perceive various functional capabilities, features, and quality characteristics into the QFD tool; the perception data can be used as weighting factors.

In collecting customer requirements for a product, the marketer should capture the customer's priorities with respect to those required. These priorities are recorded in the "relative importance" column of the QFD tool. These priorities differ from the "perceptual weightings" in the following ways:

- Relative importance is focused on measuring the degree of importance customer's place on the benefits expected with the satisfying of a given requirement.
- Perceptual weighting is focused on measuring how positively or negatively the customer perceives a given functional solution to the needing a given requirement.

Once the requirements or technical characteristics of a product have been defined, a more complete set of specifications must be prepared. This set of specifications is then used as the basis for the development of various product concepts.

Quality Functional Deployment Framework

Developed in the late 1960s by Yoji Akao and Dr. Shigeru Mizuno, the Quality Functional Deployment method provides a structured methodology for defining, organizing, and understanding customer needs and expectations, and for translating this information into product/service characteristics and requirements. One of the goals of the QFD process is to identify the CTQ characteristics necessary for customer satisfaction. The second goal of the QFD process is to ensure that these CTQs are propagated throughout the product/service life cycle planning (Fig. 14.3).

To achieve the second goal, the QFD process incorporates five domains: the customer domain, the functional domain, the design domain, the process domain, and the quality assurance domain. To achieve the primary goal, the QFD process utilized a matrix analysis tool called "The House of Quality."

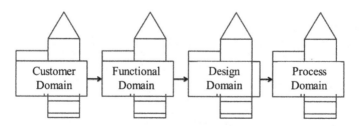

Fig. 14.3 QFD domains

The House of Quality

The House of Quality has twelve sections: objective (1), customer requirements (2), customer importance ratings (3), product requirements and technical characteristics (4), the interactions matrix (5), relationship matrix (6), competitive analysis (7), perception of quality ratings (8), technical evaluation (9), target values (10), technical difficulty (11), and importance rating (12). This structure is illustrated in Fig. 14.4.

In the objective (1) section of the House of Quality, the specified goal(s) should espouse the expected benefits that the product, or service, will provide the customer. This statement will help planners to properly analyze each functional option's ability to deliver the expected benefits while satisfying as many customer requirements as possible.

In the customer requirements (2) section, the customer's needs, expectations, and usage requirements for the product, or service, are captured. The term "Voice of the Customer" is often used in referring to this section of the House of Quality. The requirements captured by the Voice of the Customer can be collected in a variety of ways: direct interviews, surveys, focus groups, customer specifications, observations, historical warranty, and field service data, etc.

The best approach to collecting the Voice of the Customer depends upon the target market for the product, or service, being designed. Typically, in broad

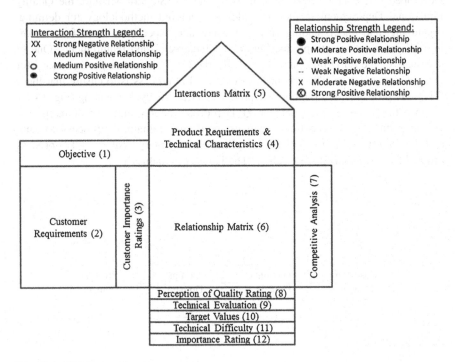

Fig. 14.4 QFD house of quality

markets, like those for computers or banking, the customer base is diverse and segmented. Likewise, in narrow markets, as with NASCAR tires or auto loans, customer requirements are often specific and frequently documented. In determining how to gather customer requirements for a specific product or service, it must be recognized that customers will usually only express a fraction of the requirements that they actually expect. Many other requirements are commonly understood, or may not even be realized by the customer, but nonetheless will lead to high levels of satisfaction. More often than not, the unspoken requirements are the most important requirements for driving customer satisfaction.

There are several types of requirements that customers may express. There are functional requirements that express how the product or service should perform. There are non-functional requirements that address cosmetic or convenience features of the product or service. There are systems requirements that define how the customer wishes to interact with the product or service. There may also be implementation requirements that define how the customer wishes to use the product or experience the service. The use of an affinity diagram can help in the categorizing of the various requirements.

The customer importance ratings (3) section defines the degree of importance that customers place on the various requirements specified in the Voice of the Customer. One easy method for helping customer determine their actual levels of importance is to give them a $100 and tell them to distribute the money between the various requirements according to the degree of value they perceive for each requirement. Having the same level of importance for multiple requirements is acceptable.

In the product requirements and technical characteristics (4) section, planners will list various product requirements and technical characteristics that can be used in responding to customer requirements. This list should be keep manageable; no more than twenty or thirty items. In determining this list, planners should look at technical evaluations of prior generation products and other competitive products. This evaluation should be based upon defined product requirements or technical characteristics. Sources of information for the evaluation can come from benchmarking, warranty and service data, third-party evaluations, etc. One of the principal outputs of this evaluation should be a set of preliminary target values for each product requirement or technical characteristic. The more successful items should be noted in Sect. 14.4, and the competitor that offered the appropriate solution should be noted in the technical evaluation (9) section.

The items listed in this section will not only provide basic functional solutions to customer requirements, but can provide innovative solutions that could "WOW" the customer. It must be understood that there are usually multiple ways of satisfying any given requirement. Some ways are technically superior to others, while others are more economically feasible. Regardless of the firm's capabilities, planners need to evaluate all functional capabilities that are technically and economically feasible for satisfying the customer's requirements. The more innovative and creative the solution, typically the more delighted the customer's response.

Irrespective of which functional options are selected, if the customer does not perceive them to be quality solutions to their needs, they will be unhappy. To help planners factor this criterion into their analysis, a tenth section has been added to the traditional matrix. The perception of quality (10) section is on a scale ranging from 0 to 2; with 1 indicating a neutral perception. Perception scores of less than "1" indicate a poor perception, thus reducing the relationship scoring developed in Sect. 14.6. Perception scores greater than "1" indicate a positive perception and will increase the relationship scoring. It must be noted that these perception scores are relative to the abilities of other functional options that could also satisfy customer requirements. The intent is to provide functional options that may not necessarily satisfy a given set of customer requirements as well as another option, but is perceived by the customer as being a higher-quality solution and therefore having a higher priority in the design domain.

The factors in the interactions matrix (5) of the House of Quality deal with the degree to which one product requirement and technical characteristic will either enhance or retard the capabilities of other requirements or characteristics. Typically, symbols with numerical values as used in the matric and defined in a legend.

In the relationship matrix (6), planners will analyze the relationships between customer requirements and the product requirement and technical characteristics. Outcomes of this analysis are used to finalize the product/service development strategy and production plans, finalize target values, and help focus planning on the critical few CTQ's.

A legend defining the strength of these relationships is developed, and the symbols relating to those strengths are used to mark which the relationship between the various product requirement and technical characteristics and specific customer requirements. The typical minimum number of strength symbols is strong, medium, and weak relationships. Weighing factors for these symbols need to be assigned. However, there are specific rules for the assigning of values to these relationships; the typical weighting factors used are: 9, 3, and 1, respectively. These numeric values are used in establishing technical priorities. The design priority section (9) of the House of Quality is calculated by multiplying the weighting factor of a given relationship, by the customer importance rating (3), by the perception of quality rating (10), and adding the products from each relationship in that column.

In the competitive domain (7), prior generation products can be evaluated against competitive products with respect to the customer requirements. This information can help designers better understand current strengths and weaknesses of various designs relative to the competition. This understanding could lead to breakthrough opportunities that would allow future designs to exceed competitor capabilities and even delight the customer.

The technical difficulty (11) associated with creating/implementing a given product requirement and technical characteristics is rate on a 1–5 scale with 5 meaning very difficult and risky. In determining this rating, planners must consider technical maturity of processes that are available, the technical qualifications of the firm's workforce, the company's manufacturing capabilities and the capabilities of its suppliers and subcontractors, costs and schedule.

The customer domain of the QFD process is developed during phase 1 of the AP/SQ&C planning. There are several objectives that must be satisfied in this domain: (1) to identify all customer requirements for the product or service being designed, (2) to establish the customer's rankings (e.g., priorities) for those requirements, (3) to identify the functional capabilities necessary to satisfy those requirements, (4) to prioritize functional capabilities that were identified based upon the strength of their respective abilities to satisfy customer requirements, (5) to analyze competitive opportunities, and (6) to establish critical characteristics' target values (Fig. 14.5).

The design domain is used to determine critical processes and process flows. By structuring the matrix, the critical part/assembly characteristics and target values that will drive the identification of production equipment requirements and establish critical process parameters are identified in the functional domain.

The process domain is used to establish process control methods, and associated inspection and test methods and parameters. By structuring the matrix, the

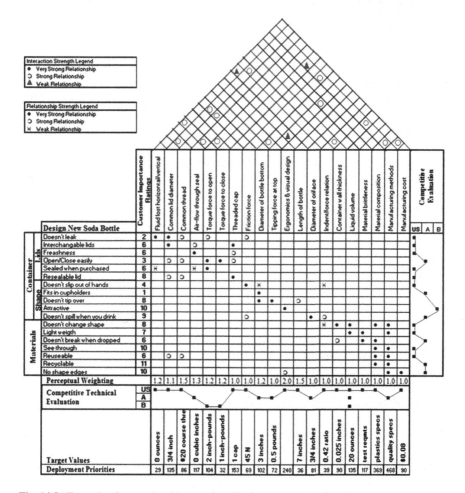

Fig. 14.5 Example of a completed house of quality in the customer domain

production equipment requirements and critical process parameters in the design domain that will drive the identification of quality control and process monitoring methods and processes are recognized.

Identifying Competitive Products

The analysis of competitive products is an essential part of establishing target specifications. Because of differing competitive strategies, competitors will have different perspectives on customer needs and how to solve them. As a result, competitive products may exhibit successful design attributes that designers should seek to emulate by incorporating the respective information into the new product's target specifications. Alternatively, competitive products can also exhibit gaps, or attributes that fail to provide customer satisfaction. Overall, a complete analysis of competitors and their products can provide designers with a starting point from which to improve, thus eliminating the need to re-invent the wheel.

14.2.2.3 Generating the Product Concept

There are numerous methods for solving any given problem, or to accomplish a specified goal. The objective of the design process is to develop the optimal design (e.g., the design the well-being in the most revenues) that is technically feasible and cost-efficient, as well as best meets customer requirements as defined by the target specifications. To accomplish this objective, designers must focus on the core product idea, while maintaining an awareness of how customers will perceive the proposed product solution.

From the set of specifications that define capture the customer's requirements, and the product's technical characteristics, various product concepts can be generated. Activities such as benchmarking, brainstorming, and research and development can provide additional information for product concepts. A concept selection matrix is a useful tool for developing product concepts.

Once the development of candidate designs is completed, a more formalized review by the team and select executives will evaluate the alternatives by looking at the trade-offs between quality attributes and their impact on the desired value proposition for the product offering. Customers may also be polled for their inputs in the selection process. At this stage of the development, the prospective design provides an overview concept of appearance and functionality, as opposed to in-depth designs and engineering.

14.2.2.4 Refining Product Specifications

Once a final product concept has been selected, designers and engineering will refine the specifications, making trade-offs between technical feasibility, expected service life of the product, projected selling price, and any financial constraints that management has placed on the development process.

14.2.2.5 Performing the Economic Analysis

The economic analysis provides management important information regarding the economic implications associated with projected development expenses, manufacturing costs, and selling prices over the life of the product.

14.2.2.6 Finalizing the Development Plan

This is the last step in the product conceptualization process. The development plan provides details on the remaining activities of the project, the necessary resources and expenses, and a schedule.

14.2.3 The Design Stage

Design is a planning process focused upon the development of a product's concept and the associated specifications that establishes the foundation of the production of a given product. This process is based upon striking a balance between functionality, quality, and cost, while maintaining the desired value proposition. For a product to be successful, it is important that both the end user and the client are considered throughout the process. There are several questions that must be considered in this design exercise:

- What is the basic goal of the product's value proposition?
- Who is the intended end user?
- What types of problems could these users experience?
- How could these users' use and/or misuse the product?
- How does the user know when he is using the product correctly?
- How can misusage be avoided?
- Does the product engage the end user?

There are several factors that will affect the overall development of the product's design. The relative importance of each factor will be governed by the desired characteristics of the product. These factors must be considered throughout the design stage of the product's development. The principal factors are the product's quality dimensions (e.g., aesthetics, conformance to specifications, convenience, creativity, durability, ergonomics, features, performance, reliability, security, and serviceability), the legal environment, company constraints, customer requirements, length of service life, testing requirements, manufacturing requirements, cost, and transportation requirements.

For a product design to be successful, it must affect the customer at several levels of their cognitive and emotional systems: visceral, behavioral, and reflective. Take the Harley-Davidson Motorcycle: its engine design is probable the most successful ever. For a motorcycle enthusiast, the sight of a Harley often engenders an immediate and positive visceral response, which triggers reflections of that V-Twin powerfully banging away between their knees and the welling, loping storm of exhaust noise emerging from its tailpipes as they open up the throttle, coupled with the feel of gritting their teeth as they go rumbling down some windy parkway, feeling as if there is asphalt being flung in their wake. As they reflect more on past experiences, they start to recall the negatives associated with those long road trips, the sweat and toil, back pain, cramped muscles, and frequent gas stops. They also start remembering the dangers of riding in bad weather and heavy traffic, and the times they had been left stranded far from home and the expense of maintenance. Thus, the idea of owning another motorcycle might start to look more than a little impractical.

Because of the power of our initial emotions take time to fade, the negatives associated with our memories do not immediately overcome the positive effects which are generated by the sight of the Harley. This type of emotional conflict is common with most designs. We interpret our past experiences at several levels; unfortunately, what is appealing to one person is not necessarily appealing to another.

A superior design will excel at all cognitive and emotional levels. At the visceral level, it will affect people at their pre-conscious state, where appearance matters and first impressions get formed. At the behavioral level, it engages with the actual usage of the product, affecting our assessment of the product's functional performance and usability. At the reflective level, it will affect our consciousness where our higher order feelings, emotions, and cognition reside, and where our thoughts and emotions are reconciled.

In applying these three levels of our cognitive and emotional systems to actual designs, we can map them to product characteristics as follows. Visceral design elements apply to the aesthetics, or appearance, of the product. The behavioral design elements apply to those features and functionality that affect the effectiveness of the product and the user's pleasure with actually using the product. The reflective design elements address the user's self-image, personal satisfaction, and memories. As we discussed in previous chapters, user perceptions are affected by many factors, so the possibility of designing a product that will satisfy everyone is extremely small. But when a product can positively affect everyone, it is a block buster success.

The design stage is a multi-phased process involving the product's functional design, detailed design, packaging designs, testing, analysis, and design reviews. Figure 14.6 illustrates the process flow of the activities in this stage.

14.2.3.1 Functional Design

The functional design phase of the design process is focused on specifying the functional characteristics of the product, rather than the aesthetics, or the number of components in the product. Functional design starts with a clear statement of what the product is supposed to do. Consideration of characteristics/capabilities that are not intrinsic to the base concept of the product being designed can improve the customer's perception of value. As an example, adding a light to indicate

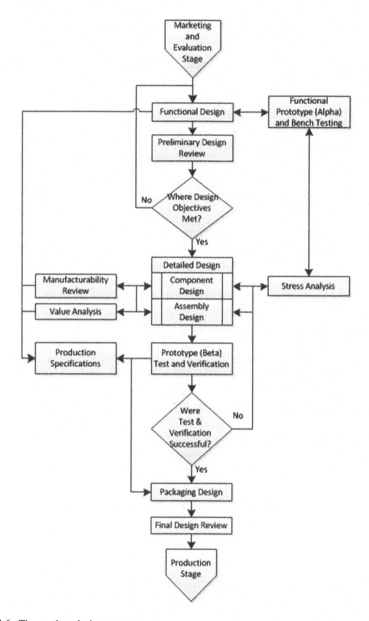

Fig. 14.6 The product design process

that the product is plugged in, using sound to indicate an inappropriate usage, or problem with the product.

Usually, there are multiple ways that a given functional result can be achieved. However, there are basic elements that need to be taken into account in order to optimize the product's intended function. When there are trade-offs to be made, the design priorities are reflected in the deployment section of the QFD tool. As a minimum, functional specifications should address the following:

- The product's performance objectives, operating conditions, and requirements for reliability and maintainability.
- Relative technological constraints.
- Safety and regulatory constraints.
- Manufacturing and installation constraints.
- Requirements for testing and development.

14.2.3.2 Prototyping

The purpose of prototyping and bench testing is to optimize the performance of the product. The bench test model provides a platform for comparison of competing versions of the product by measuring primary performance attributes without concern for form.

The alpha prototype is an experimental model that illustrates the product concept relative to technical specifications and elicits feedback. The alpha model is typically produced using non-production materials.

The beta prototype is built using production quality materials and methods. The purpose of the beta model is to verify performance and reliability of the actual product. Both the alpha and the beta prototypes can also be used to discover possible failure modes.

A major benefit of prototyping is that it is a tool for improving the design of the product. Should customer feedback be need before market introduction, a gamma prototypes can be used. The gamma model would also include the consumable items and services included in the value proposition.

Prototype testing is generally focused upon stressing the product up to and beyond conditions expected to exist in the field. These tests may even push the product to its failure points to confirm where those points exist.

14.2.3.3 Detailed Design

Once the functional concepts and specifications have been validated and accepted, product engineering will start producing the detailed drawing of components and assemblies used in the final product, and the overall configuration of the final product. The purpose of a detailed design is to transform the excepted product concept, functional specifications, and technical requirements into a set of manufacturing documents.

The first step of this process is to disaggregate the product concept into its logical assemblies and components. To help in this exercise, the quality component deployment (QCD) tool shown in Fig. 14.7 provides a framework for

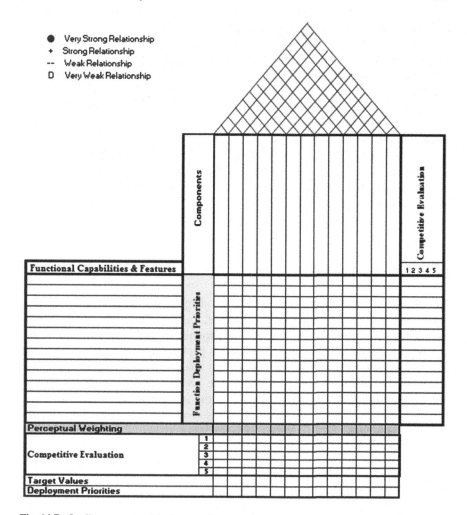

Fig. 14.7 Quality component deployment framework

the rationalization of assemblies and components with respect to their ability of deliver the necessary functional characteristics. To insure that the assemblies and components selected help to achieve the perceived quality desired, customers should be surveyed and their perceptions recorded as a weighting in the tool.

Next, the design and/or selection of the best fitting component parts and subassemblies are integrated into the product's final configuration. Documents produced by this process are as follows:

- Detailed component drawings and material specifications.
- Detailed assembly drawings and processing specifications.
- A bill of materials.

14.2.3.4 Design Verification and Testing

The test and verification stage involves the development of both functional and performance prototypes, test procedures, test equipment, and the actual testing and documenting and reporting of test results. The objective of design verification and testing is to assure that the product meetings all of the specifications the define customer requirements for satisfaction.

The beginning of this process starts with the translation of design specifications into a test plan. Tests can occur at multiple points within the design, production, and delivery processes. In preparing for verification testing, the best approach to conducting the verification must be determined. Then, measurement methods must be defined, along with the identification of the necessary facilities, tools, and instruments for the conducting of the testing.

14.2.3.5 Packaging Design

The design of product packaging is focused upon the enclosing or protecting of the product for distribution, storage, sale, and use. The objective of packaging is to protect the quality of the product during transport, warehousing, logistics, sale, and end use. A product's packaging is often the customer's first visual experience with the product. It therefore can have a significant effect on the customer's perceptions of the product. The process of designing the packaging of a product starts with the identification of requirements: structural design, shelf life, quality assurance, logistics, legal, regulatory, environmental, etc.

14.3 Commercialization

The actual launch of a new product is the final stage of product development process. This launching process is commonly called "commercialization." In several respects, commercialization overlaps the development process to the point where they are almost synonymous. Commercialization is a multiple phase process, ranging from the concept through production, distribution, marketing, and customer support. As a strategy, commercialization requires that the company develop a marketing plan, determine how the product will be supplied to the market and anticipate the probable barriers to success. Included in both, the development of the product concept and in the marketing plan should be an understanding of how the customer will perceive the product and how to enhance that perception.

14.4 Implications to the Customer

From the traditional competitive perspective of manager's, being successful is a matter of offering consumers greater value. Thus, quality versus price is the main trade-off to be made. Their logic for this assumption is linked with Deming's chain reaction model which emphasis that when the company improves its product quality, customers will perceive a greater value that will generally result in either higher market demand, or if the company chooses to restrict the supply of the product, then it can increase its pricing, both of which lead to the obtainment of higher market share. Thus, the company can stay in business, providing more and more jobs. In trying to leverage this assumed behavior against customers, Garvin points out that "if managers believe that perceptions and perhaps consumer purchasing decisions are positively correlated with price, they may set higher prices in order to imply higher product quality."

Depending upon the market, designers do not need to pursue all of the quality dimensions. Instead, they can emphasize a select few to focus on: thus, possible enabling the company to capture a dominant share of a niche market. In fact, as the diversity of perspectives in a given market space increases, this approach is most likely the best one.

14.5 Product Improvement Methodologies

It is during the design stage that most of a products life cycle costs, functionality, and overall benefits that affect customer satisfaction are committed. The design of a product is not just based on practical or innovative design practices. Marketable design must be manufacturable at a cost that is competitive as well. In other words, design what the customer will value, but do so within the capabilities of the company's various processes. To accomplish this goal, development personnel will need to determine that critical characteristics are within the processes capabilities of the company or else their processes will need to be improved. The primary assumption is that the company has stable processes in place.

Design reviews should be a part of any formal evaluation process that is conducted during the development cycle of a product, or service. These reviews ensure that all requirements and concepts are satisfactorily addressed prior to proceeding into production. The reviews also evaluated whether or not issues were properly understood, risks were appropriately managed, and a good business case exists for moving forward. There are two basic types of design reviews: a function-based review and a value-based review.

14.5.1   Function-Based Reviews

Function-based reviews primarily focus on designing products in such a way that they are easy to manufacture. This type of review not only focuses on the design aspect of a product, part or assembly but also on its producibility. Function-based reviews have several names, such as design for manufacturability (DFM), design for manufacturing, design for assembly, and design for Six Sigma (DFSS).

The manufacturability review is very important. The objective of this review is to insure that the features of the detailed components, including the tolerances and materials, are compatible with the firm's existing manufacturing processes and technology, thus ensuring that quality is an integral part of the product design, through the optimal selection of parts and proper integration of parts, for minimum interaction problems. Other objectives of this review include the simplification of the designs as much as possible without compromising the intended functionality, and cost reduction, the optimization of all manufacturing functions. These include fabrication, assembly, test, procurement, shipping, delivery, service, and repair. The manufacturer has to assure that the best possible combination of cost savings, quality, reliability, regulatory compliance, safety, time to market, and customer satisfaction is achieved.

The process of reviewing designs for manufacturability tends to encourage the standardization of parts, maximum use of purchased parts, modular design, and standard design features. The general guidelines for this type of design review are as follows:

- Keep it simple: fewer parts and less complexity are better.
- Understand the manufacturing process, including all problems with past and current products with respect to manufacturability, quality, repairability, serviceability, and performance. It is also important to consider the effects on operations that customers notice. These include inventory management, order-handling, incoming inspection, and operating costs as well as operations that affect suppliers, such as downtime, tooling, warranty costs, and process yield rates.
- Design components and subassemblies for ease of fabrication, and material processing.
- Ensure that all specifications are using proven processes, such as welding, casting, forging, extruding, forming, stamping, turning, milling, grinding, and plastic molding, and cost-effective technologies and methodologies.
- Avoid use of mirror image parts.
- Use symmetry in the design wherever possible; otherwise, make the parts asymmetrical.
- Tooling and fixturing.
- Use geometric tolerancing.
- Optimize the robustness of design.
- Use components and materials from reliable sources with proven and consistent quality.
- Minimize processing setups.

- Minimize the number of cutting tool operations.
- Sustainability.
- Materials.
- Fool proofing assembly designs.

14.5.2 Value-Based Reviews

Value-based reviews are commonly called value analysis (VA), or value engineering (VE). This type of review is a function-oriented, structured, analysis process that is focused on: (1) identifying and distinguishing between incurred and inherent costs, (2) minimizing costs, and (3) improving product performance. This analysis examines the materials, processes, work methods, and both information and material flows used in the production of a product or service. The intended benefits include reduced material usage and costs, reduced waste, improved profit margins, increased customer satisfaction, and increased employee morale.

The review process compares the customer's required functionality for a product, at its lowest cost, to specific performance and reliability requirements. The objective of the process is to identify and eliminate product and service features that do not add value to the customer.

Traditionally, VA has been focused upon cost reduction as opposite to quality enhancement; thus, VA can indirectly improve the customer's perception of a product by increasing its value. Going forward, VA should also evaluate the impact of how various features and processes affect the customer's perception of the product or service. Ultimately, the goal of VA activities is to offer a product or service that the customers will value at the lowest optimal cost of production.

The potential benefits of VA are as follows:

- Reduced project costs.
- Decreased operation and maintenance costs.
- Less paperwork.
- Simpler procedures.
- Improved project schedules.
- Less waste.
- Increased procurement efficiency.
- More effective use of resources.
- Development of innovative solutions.

The knowledge and skills required to effectively perform VA are as follows:

- Technical facilitation and group leadership.
- Knowledge of basic economic principles.
- Knowledge of scheduling and sequencing of design and construction processes.
- Ability to communicate and interact with multiple disciplines.
- Understanding and appreciation of interdisciplinary issues.
- Creativity and the ability to think outside the box.

14.5.2.1 A Basic Approach to Value Analysis

An effective VA generally follows a work plan and procedures. The work plan should be structured to obtain maximum effect from the effort expended to perform the analysis. By using a formal plan, the VA team assists in a number of ways as follows:

- The plan provides an organized approach to analyzing a project by identifying areas in which value improvement may be possible and selecting alternatives that minimize costs while maximizing quality.
- A work plan encourages the team to think creatively and to look beyond the use of common or standard approaches.
- A well-crafted work plan emphasizes total ownership costs for a facility, rather than just initial capital costs.
- A work plan leads the VA team to developing a concise appreciation of the purposes and functions of the facility.

14.5.2.2 Cost Versus Value

There are two basic considerations associated with improving the value of a product or service: usage (i.e., use value) and the source of value (i.e., esteem value). Consider the difference between a gold-plated, wood-trimmed, ballpoint pen and a disposable pen. Both pens fulfill the same usage: to apply ink to paper in a controlled fashion. Both pens achieve their use value equally, but the price points for these two pens are very different. The difference is because of their respective esteem values.

Managers need to understand the nature of costs both for the factory and for any given product. Often, then is little or no direct relationship between a product's cost at the factory and customer value in usage and esteem. Typically, 80 % of a product's manufacturing costs are determined prior to the release of engineering drawing to manufacturing. Because of this, it is essential that periodic reviews occur during the design process in order to recover any "avoidable" costs while maintaining the "use value" to the customer.

Traditional thinking captures production costs in three basic categories: the cost of purchased parts and materials, the costs of direct labor, and the cost of manufacturing overheads. Most cost reduction efforts typically focus solely on direct labor. This approach misses the greater opportunities presented by the other two cost areas with the removal of unnecessary materials and overhead costs. In looking at the "total" costs of a product, management needs to understand the way costs accumulate in the factory, as well as the relationship between these costs and value generation. Thus, new sources of cost reduction are found in the cost of manufacturing, the cost of assembly, the cost of poor quality, and the cost of warranties.

The main focus on costs is to reduce it as much as possible without negatively affecting the functionality, quality, reliability, maintainability, or other benefits expected by the customer. The goal is to accomplish the cost reduction without having to re-engineer the product or service design. Thus, there are three simple rules for the conduct of a VA analysis:

- Cost reductions shall not compromise the quality or reliability of the product or service.
- Cost reductions shall not compromise the saleability of the product or service.
- Cost reductions shall not increase to the cost of ownership to the customer.

14.5.2.3 Issues Driving the Value Analysis Process

VA is a structure approach for the review and cost reduction analysis of product and service systems designs. The reasons driving a VA analysis can be divided into two basic sources: those that lie within the business and those that stimulated by market activity. Figures 14.8 and 14.9 illustrate an analysis of a mechanical pencil using the VA process.

Design-Related Issues

- Designers may be unaware of "best practices."
- Designers may not understand or beware of current technical capabilities in manufacturing.
- Designers might not be thinking through the implication associated with traditional thinking, or customary practices.
- Designers may have taken short cuts to meet deadlines, or political pressure.

Internal, Non-design-Related Issues

- Products with known problems.
- Customer demands.
- Safety.
- Regulatory compliance.
- Product profit margin improvement actions.
- Corrective actions to redress problems.

Market Sources

- Pricing.
- E-commerce.
- Product complexity.
- Quality/regulatory compliance.
- New technology and materials.
- Environmental/social pressures.

Best Practices

- Gain approval of upper management.
- Enlist a senior management champion.
- Select the product
  - Define and communicate the objectives of the project.

- Select a team
  - Select an operational leader.
  - Colocation team members.
  - Train team members.

- Establish the reporting procedure.
- Present VA concept and objectives to management.
- Regular communications.

14.5.2.4 Implementing a Value Analysis Review

There are typically seven phases associated with the implementation of a VA review.

PHASE 1: Start–up Phase

Once it has been determined that the costs associated with a specific product or process is not the fault of any individual, then the VA process can be utilized to

Value Analysis of a Mechanical Pencil

Customer Requirements / Functionality	Functional Importance	Lead	Eraser	Body	Finish	Band
				Mechanisms		
Make Marks	30	◉ 150				
Remove Marks	20		◉ 100			
Prevent Smudges	15	○ 45		○ 45		
Support Lead	5			◉ 25		
Aesthetics	10			○ 30	○ 30	△ 10
Ergonomics	20			◉ 100	△ 20	
Column Weight	535	195	100	200	30	10
Perceptional Effect		1.00	1.00	1.10	1.20	0.90
Adjusted Weight	560	195	100	220	36	9
Priority Weight	1.00	0.348	0.179	0.393	0.064	0.016
Target Cost	$3.10	$1.13	$0.58	$1.16	$0.18	$0.06
Actual Cost	$3.23	$1.33	$0.48	$1.04	$0.11	$0.28

Fig. 14.8 Value analysis matrix

Value Analysis of a Mechanical Pencil

Customer Requirements / Functionality	Functional Importance	Mechanisms Lead	Eraser	Body	Finish	Band
Make Marks	30	150				
Remove Marks	20					
Prevent Smudges	15	45		45		
Support Lead	5			25		
Aesthetics	10			30	30	10
Ergonomics	20			100	20	
Column Weight	535	195	100	200	30	10
Perceptional Effect		1.00	1.00	1.10	1.20	0.90
Adjusted Weight	560	195	100	220	36	9
Priority Weight	1.00	0.348	0.17	0.393	0.064	0.016
Target Cost		$3.10	$1.13	$0.5	$1.16	$0.18	$0.06
Actual Cost		$3.23	$1.33	$0.4	$1.04	$0.11	$0.28

Alternative Designs for Mechanical Pencil Body

Customer Requirements / Functionality	Functional Importance	Alternative #1	Alternative #2	Alternative #3	Alternative #4	Alternative #5
Make Marks	30					
Remove Marks	20					
Prevent Smudges	15	45	60	45	45	45
Support Lead	5	25	25	25	25	15
Aesthetics	10	30	50	50	30	30
Ergonomics	20	100	60	60	60	60
Column Weight		200	195	180	160	150
Perceptional Effect		1.10	1.00	1.10	1.20	0.90
Adjusted Weight		220	195	198	192	135
Priority Weight		0.393	0.393	0.393	0.393	0.393
Target Cost		$1.16	$1.16	$1.16	$1.16	$1.16
Actual Cost		$1.04	$1.01	$1.09	$1.12	$1.10

Fig. 14.9 VA alternative analysis matrix

re-establish the value of the product, or to design an effective monitoring and control system for the process.

Team Building

Next, management must select and develop a VA team. The exact composition of a VA team will depend upon the specific needs of the company and the availability of resources. Ideally, the team will be composed of stakeholders in the product design and manufacturing process from within the company and external representatives from customers and suppliers that are being affected by the costs of the poor performing products or processes. It is essential that each member possessed both technical skills associated with one or more functional areas affecting the production and support of a product or service, and good problem-solving skills. Each member must also be able to communicate well. An understanding of basic economic principles, production methods, and materials are also important.

The internal members of the VA team need to be drawn from technical, skilled, and professional staff (e.g., designers, manufacturing engineers, purchasing specialists, and operational staff). These members tend to possess a high degree of understanding of the product, its production processes, and the failures associated with them from their functional perspectives. These members also understand which solutions would be must feasible in overcoming these difficulties. In addition to the technical specialist, individuals with only a general overall knowledge or skill set should be included in the team's composition because their simple basic questions can stir thought provoking discussions, which in turn can result in breakthrough solutions.

The addition of external members not only adds more information to the decision-making process of VA, and it also provides an efficient and effective method for greater customer and supplier engagement. With the increasing levels of competitions in virtually every market, management has been forced to adopted new operational philosophies and organizational structures in or to remain competitive. Examples of this can be seen in shift from vertically integrated organizations toward flexible, value chain oriented enterprises. It can also be seen in practices such as outsourcing.

The inclusion of customers in the VA process brings greater insights into the customers' real intentions for the products and services they consume. Furthermore, customer information is a key ingredient to the success of any VA activity. Customer inclusion also promotes their understanding of the VA process and the benefits that can be derived from it as opposed to their simply using the process to drive the price reduction of the products or services they use.

The addition of suppliers and subcontractors to a VA process brings several benefits to the team. First, suppliers and subcontractors know more about their products and services than most designers. Not only are they familiar with the technical aspects of the product, but they are also familiar with their effectiveness in various applications. Because of these strengths, suppliers tend to make better design decisions with respect to cost and functionality where there products and services are used. The inclusion of suppliers and subcontractors also promotes concurrent engineering, thus collapsing project development times.

After the review team has been finalized, best practices suggest that the team as a whole take a tour of the facilities used in the development and production of the product or service to familiarize themselves with processes of design and manufacturing. In addition, the team should visit select customers, touring their facilities and interviewing key personnel to better understand the customer's current problems and future needs. Finally, team building exercises will help promote comfort and familiarity between team members, making them more effective.

Product/Service Selection

The selection of product or service designs to be reviewed should be based upon an established set of criteria and attributes. These criteria and attributes often include, but are not limited to, the following items:

- Known problems: For established products or services, significant number of customer complaints or costly warranty returns are a good indication that there are problems with the offering that are compromising the customers' ability to perceive its value. For new products or services with design, issues can cause high levels of scrap and/or rework costs, as well as internal process changes before the offering reaches a consistence level of quality that meets customers' requirements.

- Significant changes in forecasted sales volumes: There is a high degree of inter-dependence between a product's design, the manufacturing process' design, and the companies marketing strategies. When these three items are in sync, an optimal cost structure for production and delivery can be achieved. Whenever market demands are significantly different from expected demand levels, man-ufacturing efficiencies will drop, and costs will increase. Changing production processes can be prohibitively expensive and may even drive product design changes. VA activities typically focus on high-volume products, as they offer the greatest returns and frequently have the greatest impact on company profits.
- Below average profit margins: As with the problem of changing sales volumes, underperforming products are either the result of a mismatch between product design, manufacturing process design, and marketing strategies, or a very poor design to begin with. In either case, products or services with poor of negative margins can have a significant impact on a company's competitiveness and must be addressed.

PHASE 2: Information Phase

It is essential that the team develops an understanding of the product or service under review, including the material and information flows through the supply chain, as well as the firm's production facilities. The first step in this phase is a thorough review of the project and its scope; the cost models used in the develop-ment of the projects cost estimates; and the development of a list of the key issues and value objectives. Then, data collection activities start with the product/service design, background, constraints, and projected costs of the project. Materials that typically provide data for this phase are as follows:

- Project and budget development documents.
- Engineering drawings.
- Specifications of the major construction elements of the project.
- Line item cost estimates.
- Definitions of the major systems and subsystems to be developed by the project, as well as details of any supporting systems and subsystems.
- Verification of available utilities.
- Details of any special requirements.
- Economic data, budget constraints, facilities, etc.
- Financial information and supporting data.
- Customer complaint and surveys.
- Useful techniques for this phase include;
- Brainstorming of project requirements, critical path, and losses at each stage of the design process.
- Development of a detailed product–process map.
- Development of process flow diagrams and quality loss charts.
- Development of process cost charts
- Benchmarking

PHASE 3: Function Analysis Phase

Functional analysis is the systematic process used to identify the key functions of a product or service, that define the benefits of the product or service, and through which value is generated. There are several steps to performing a functional analysis.

Describe the Functions

These functions, when properly executed, contribute to the salability of the product or service. The basic output of a functional analysis is a list of these functions. Functions are best described using verbs and nouns, such as boil water or pressure relief. Typically, a product or service will perform multiple functions and in describing them none should be taken for granted.

From the customer's perspective, there are two types of functions provided by any product or service: work functions and sell functions. Because the importance of some functions is not always immediately obvious, brainstorming can be a useful tool in identifying them. Techniques such as tree diagrams or function analysis system technique (FAST) can be used to group functions. The basic groups for the functions are needs and wants. The needs grouping is composed of functions that are either essential to the performance of a task or function, or fulfill a basic needs expressed by the customer. The wants grouping is made up of functions that are not necessarily essential to the task or function, but are essential to improving productivity and or acceptance, or alternatively fulfill a desired want of the user.

Rank the Functions Using Pairwise Comparison

The pairwise comparison procedure is used to rank a product's functions, or a service's benefit generating attributes. In this procedure, each function is compared by importance to every other function. The most important function of each pairing is identified and posted on a table. It is critical that the team makes a definitive decision as to which function in each pairing is the most important. Next, the team must score the degree of difference in importance between functions in each pairing: minor (1 point), medium (2 points), or major (3 points). Then, the scores for each function are added up; thus, highest score indicates the most important function. Experience has shown that usually only one or two functions emerge as being significantly more important than the others. These high importance functions should be targeted for the most attention in subsequent VA activities.

PHASE 4: Creativity Phase

This stage of the VA process is concerned with the development of alternative solutions to achieving the basic functions of the product or service. The goal is to find those alternatives that improve the customer's perceived value, in a more cost-effective manner and/or reduces risks.

Brainstorming is one technique that can help the team in coming up with ideas for these alternative solutions. Through this collaborative process, ideas are generated.

PHASE 5: Evaluation Phase

Once the team has concluded generating ideas for the value improvements of the product or service; they need to be evaluated for their feasibility. Each idea must be assessed and rated for its ability to meet project needs, budget constraints, and other key objectives. These selected ideas are pasts on for further consideration.

PHASE 6: Development Phase

In this phase, the VA team researches each of the selected ideas and prepares descriptions, sketches, cost estimates, and supporting recommendations.

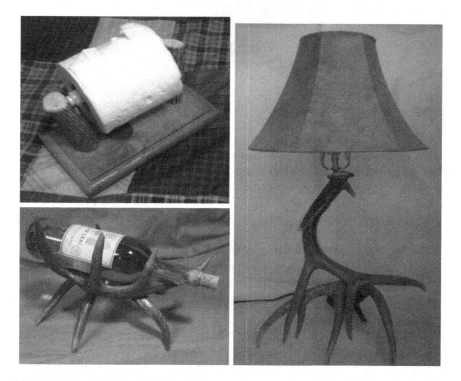

Fig. 14.10   The Antler Shed's product line

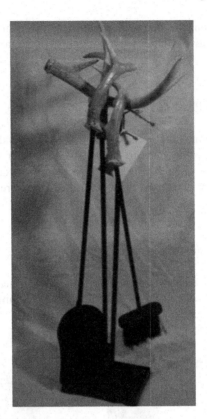

Fig. 14.11   Fireplace set #1

PHASE 7: Presentation Phase

At the conclusion of the VA process, the proposals for the selected best alternatives are presented to the product manager, designers, and other interested management. This final report or presentation summaries the results of the VA process
to date, emphasizing innovative recommendations that avoid unnecessary costs
and are tailored to meet the project's needs, objectives, budget, and key value
objectives.

The VA team and the final report should avoid selecting winners or losers,
instead focusing only on making recommendations. The project manager, design,
and management are responsible for the final decisions and their implementation.
The presentation/report will usually contain the following:

- A scope statement for the analysis.
- A description of the business conditions.
- The justifications driving the enhancement of the product or service.
- A description of the current costs of the product or service, and the failures in
  the design process that represent the costs of poor design.

- An analysis of the product and its functionality for the customer, or the service and its benefits.
- The proposed changes and the commercialization reason supporting the changes.
- The cost/benefits analysis.
- Lifecycle cost projections.
- A list of issues that could not be resolved by the analysis.
- An appendix of supporting data (Figs. 14.10 and 14.11).

14.6 CASE: The Antler Shed

The Antler Shed is a small privately held company that manufactures home decor items from the horns of antlered animals. Their current line of products, shown in Fig. 14.12, consists of several table lamp designs, wine bottle holders, steak flippers, fire ring pokers, fireplace sets and more. Each piece is handcrafted.

The primary differentiator between similar items in the Antler Shed's product line is typically the number of horns used in the assembly, along with the size and quality

Fig. 14.12 Fireplace set #2

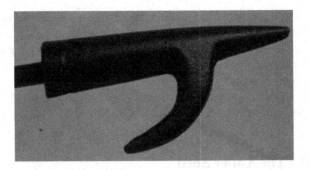

Fig. 14.13 Fireplace set #1 Log Poker tip

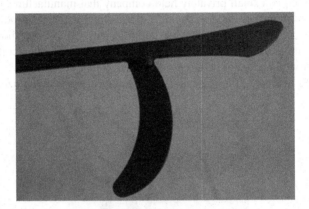

Fig. 14.14 Fireplace set #2 Log Poker tip

grading of the horns. Though most of the products are primarily handcrafted, some products incorporate a significant percentage of purchased components, as with the fireplace sets. These purchased items can affect the customer's perception of the sets quality. Below is an example of two fireplace sets that utilize purchased components.

Between the two fireplace sets shown in Figs. 14.13 and 14.14, fireplace set #1 is often perceived as being a higher-quality set than fireplace set #2, because of the additional horn used to hold the implements, as opposed to the other set that utilizes a pre-manufactured bracket. It is not until customer starts using the first set that he realizes that the top piece does not hold the implements securely. The pre-manufactured bracket resolves the functionality problem, but at the expense of appearance and perceived quality. The alternative solution of selecting a horn with more tangs, or the correct curvature, would maintain the perceived quality and appearance, but at additional cost.

The second perception problem between the two fireplace sets is the Log Poker. Both implements utilize a manufactured tip. Fireplace set #1, Fig. 14.13, uses a cast tip that appears more durable and functional, while the tip for set #2, Fig. 14.14, appears lighter and of lower quality. In actuality, both tips perform equally and cost approximately the same.

Chapter 15
The Dimensions of Service Quality

There are three aspects to services that distinguish it from physical goods: intangibility, heterogeneity, and inseparability. Because of intangibility, services cannot be touched or measured in any way before their purchase by the consumer. Furthermore, firms often find it hard to link customer perceptions of their service and the actual quality of the service. The intangible nature of services also makes it difficult to judge because of the inconsistency of behavior of service personnel (heterogeneity) and the consumer's active role in the consumption of the product (inseparability). The degree of variance in the delivery system exacerbates this problem; performance will vary from producer to producer, from customer to customer, and from day to day. Unlike with manufactured goods, the inseparability of services from consumption implies that its quality cannot be engineered, measured for conformance to specifications, and then delivered intact to the customer.

Customer perceptions of quality are influenced by the manner in which the service is delivered (e.g., functional quality) as well as by the outcomes (e.g., technical). A third factor that provides significant influence on customer perceptions is the physical aspects of the service, such as equipment, facilities, and the company's brand image. It is through the interactions of these factors with the customer that quality perceptions are developed. Because there is a tremendous difference between individual customers and even with the same customer from point of contact to point of contact, it is important that the service providers are responsive to the needs of each customer.

15.1 The Reliability Dimension

Reliability is measured by the degree of consistency and dependability with which the expected level of service is delivered to the customer. It means doing it right the first time. It means honoring promises. Thus, reliability implies that different interactions with the customer invariably lead to the same desirable result.

© Springer-Verlag London 2015

G.N. Kenyon and K.C. Sen, *The Perception of Quality*,

DOI 10.1007/978-1-4471-6627-6_15

When customers engage a service, they have formed a mental list of criteria against which they judge the success of the service. This list of criteria basically defines the quality attributes of the service and the expected benefits associated with the varying degrees to which these attributes conform to their definitions. These definitions are the customer's credence perceptions of the service. By defining each instance of a service delivery as a transaction, the service is fulfilled when these transactions meet or exceed these quality criteria. The more the attributes of the service that meet their respective quality criteria and/or the more often the service is deemed satisfactory or excellent, the more reliable customer will view the service.

15.1.1 Attributes of Reliability

Accuracy involves doing things right the first time. It is often thought of as a measurement which associates the outcomes of a task with some standard of performance or specification. It implies that the task was performed within a set of accepted parameters that meet the customer's needs.

Repeatability involves being able to achieve consistent results each time a task is performed. It implies that results are process driven and not significantly affected by factors such as which employee provided the service or which day of the week it is.

15.2 The Responsiveness Dimension

While reliability encapsulates the consistency of the service offering, the service provider must also have the ability to respond to the customer's needs. Customers often gauge a company's responsive based upon the speed of the service, the service provider's sensitivity to the customer's concerns, and the service provider's level of awareness to changes in the needs of their customers. Thus, responsiveness denotes the firm's degree of flexibility and the ability to design the service offering to a customer's special needs. One such special need is in the area of post-purchase support. This need is of most concern with commercial customers.

To improve their responsiveness, companies can do the following:

- Update and/or add new service channels to address customer service issues more quickly. Examples include the updation of telephone systems, the stream-lining of automated response systems, the creation of direct extension options can route calls to appropriate representatives faster, or the addition of email, live chat, and instant messaging tools to the customer service portfolio.
- Recruit customer service personnel that have the aptitude and desire to provide customers with effective solutions to their problems on first contact. They can

also train existing and new employees to better understand the importance of customer service. Training and other developmental activities should reinforce positive attributes with ongoing opportunities. Companies must ensure that efforts are directed toward increasing responsiveness to customers' needs by being sensitive and considerate to their individual circumstances.

- Set up systems to facilitate the collection of information gained directly from your customers. Managers must be aware of the value of this information in the development of strategies for addressing similar needs across your customer base.
- Implement systems for the requesting and collection of feedback from the customer.

15.2.1 Attributes of Responsiveness

Willingness is the degree to which providers are enthusiastic about performing the service requested.

Readiness involves being prepared to deliver the service.

Timeliness is defined as performing a requested service at a suitable or opportune time. Timeliness is perceived as an important component of dependability. It also includes following up or contacting with customers when they indicate an interest in the service.

Flexibility involves the willingness and ability to modify the service offering to the specific needs of the customer.

15.3 The Assurance Dimension

The ability of the service provider to convey trust and confidence in customers is denoted as assurance. The degree with which the service employee can convey these qualities at the points of contact with customers is critical to service quality. Both proper training and effective promotional campaigns help develop this dimension.

In efforts to improve service assurance, companies have started adopting service assurance management practices such as:

- Fault and event management.
- Performance management.
- Probe monitoring.
- Quality of service management.
- Network and service testing.
- Network traffic management.
- Customer experience management.
- Service-level agreement monitoring.
- Trouble ticket management.

In practice, firms seek to identify faults in their processes and resolve them in a timely manner so as to minimize service downtime. This practice also includes proactively identifying issues and diagnosing and resolving any service malfunctions before customers are impacted. There are many drivers for service assurance adoption, with some considered the most important to be the ability to measure the performance of a service.

15.3.1 Attributes of Assurance

Credibility is earned by taking the customer's best interest to heart, and being trustworthy, believable, honest, and having the willingness to act in the best interests of the customer.

Competence is developed through acquiring the knowledge and skills necessary to perform the service requested. It implies that all employees in the organization are knowledgable and skilled at their respective tasks and that supporting activities are capable of meeting the needs of the organization.

Communication means keeping the customer informed about the service and any issues that arise with the service they requested. Effective communication is established by the following:

- Using language that the customer is comfortable with and listening to their concerns.
- The ability to explain the service itself, including the costs and benefits associated with the service.
- The ability to explain the trade-offs between service and costs.
- The ability to assure customers that problems are being handled.

Follow-ups involve checking with the customer after the service to verify their satisfaction and/or to inform them of additional services or benefits that they may be interested in.

Security relates to the degree of risk associated with transactions during the service delivery. One of the main issues in online security is the perceived lack of security on public domains and that many customers desire to retain some level of privacy and anonymity. These issues require companies to be very responsible for both customer transaction activity and personal information. The three principle aspects relate to physical safety, financial safety, and confidentiality.

15.4 The Empathy Dimension

The service provider's willingness to see the service interaction from the eyes of the customer and to design a service that caters to each individual needs is encapsulated by the dimension of empathy. Often, "empathy" depends on the intuitive knowledge that some firms have about their customer's actual needs and wants.

15.4.1 Attributes of Empathy

Understanding involves making an effort to learn who the regular customers are, learning what specific requirements each customer has, and providing them with individualized attention.

Courtesy is measured by the degree to which service personnel are polite, respectful, considerate, and friendly toward customers and each other. This includes being considerate of other people's property and being clean and neat in their personal appearance.

Convenience involves the time spend in obtaining the service from the moment of entry into the service system to the moment of exit. Factors that are associated with convenience are as follows:

- The degree of personal involvement with the service delivery process.
- The amount of time necessary to secure the service.
- Transactional activities required in securing the service, such as qualifying procedures, loan processing, checkouts, or payment processes.

Access involves the degree to which service personnel and facilities are approachable and easy to contact. Factors that are associated with access are as follows:

- The convenience of location.
- The convenience of the hours of operations.
- The convenience of the service system design and layout of the facilities.
- Having short or reasonable waiting times for receiving the service.

15.5 The Tangibles Dimension

While services are essentially intangible, many services are accompanied by tangible cues. Customers frequently use these cues to judge the quality of the service. In most cases, these cues are only found at the service delivery site. These cues are typically associated with physical facilities, tools and equipment, personnel, and fellow customers at the service site.

15.5.1 Attributes of Tangibility

Product Portfolio encompasses the range and depth of product and service offerings that are part of the service experience. With online service functions, it also includes the number of useful free services and diversity of features. Customers engaging in online transactions often seek products/services that are unavailable in local outlets. Thus, limited product selections or outdated information about offerings will negatively affect customer satisfaction.

Facilities should postively represent the intangible elements of the service with respect to their physical appearance and ambiance.

Personnel should be attired appropriately.

Tools and Equipment not only must be capable of performing the tasks required for the delivery of the service, but should also be properly maintained and logically organized.

Artifacts are the physical representations of the service, such as statements and credit cards. It is important that they must convey elements of the service theme as well as be appropriate for the task or function they are used for.

Clientele have a direct impact on the service environment. They can either have a positive or a negative effect on other customers' opinions of security and inclusiveness while receiving the service.

15.6 The Ease-of-Use Dimension

Ease of use relates to the degree of ease with which customer can navigate through the Web site, the degree of organization and structure of the Web site, and the ease of completing an online transaction. Does the content of the Web site provide easy-to-use features essential to attracting both experienced and new online customers? Site content should be concise and easy to understand. Overall, simplicity and smoothness of the whole transactional process is essential to ensure customer satisfaction.

Service Quality Dimensions:	Stakeholders			
	Investors	Employees	Customers / Consumers	Society
Reliability (accuracy, repeatability)	✓	✓	✓	?
Responsiveness (willingness, readiness, timeliness, flexibility)		✓	✓	?
Assurance (credibility, competence, communication, follow-ups, security)		?	✓	?
Empathy (understanding, courtesy, convenience, access)	✓	?	✓	
Tangibles (product portfolio, facilities, personnel, tools & equipment, artifacts, clientele)		✓	✓	
Ease-of-Use (information, function, features)	✓	✓	✓	

Fig. 15.1 Service quality dimensions visibility to the customer

15.6.1 Attributes of Ease of Use

Information relates to the degree to which instructions are clear, concise, and unambiguous.

Functionality relates to the range of operations by which the service is delivered and the degree to which they are successful in generating the expected benefits of the service. The easier the functions are to use, or to engage with, the greater the value of the perceived benefits.

Features are the various options available to the customer within each function. The customers choose options will define the benefits related to the function.

Figure 15.1 provides a mapping of the quality dimensions that define our perceptions of a service to the customer/stakeholders that are vested in the service.

Chapter 16
Service System Design and Commercialization

At its essence, a service is someone else doing your work for you; thus, everyone will have specific and semi-unique requirements and expectations for the quality of the service rendered. The design of service systems is particularly difficult because of the intangible nature of the attributes associated with services. Due to this characteristic, customers are not able to touch or easily measure the elements of the service they receive. Thus, their perceptions of quality are based on qualitative assessments of how well they believe the service has met their expectations. Because of this subjective assessment of quality, service providers must design their service delivery systems in ways that control customer expectations for the service and their perceptions of that service.

To facilitate the service provider's ability to shape customer expectations and perceptions, it is essential that they understand the nature of the attributes that define the quality in their service offering as well as the degree of relevance each attribute holds for the customer. Thus, service providers must understand the background and motivations of their target market, as well as the nature of the process through which consumers create their perceptions of quality.

With the increasingly higher degrees of global competitiveness and the number of substitute products available, services rather than the manufactured goods provides the firm with the greatest opportunity to differentiate itself from its competition. This in turn allows the firm to gain a sustainable competitive advantage. To leverage competitive opportunities, selected services are frequently bundled with manufactured products. Understanding the relationship between service attributes, product characteristics, and customer needs is imperative.

Over the last several decades, numerous advancements in communications and information technology have provided service providers with tools for collecting and analyzing a wide array of data about consumers. In fact, whole industries have been created simply for that purpose. Internet service providers data mine customer accounts and transactions, third-party marketing firms are continuously surveying customer opinions, and companies track point of sales transactions as well as collect customer data such as zip codes and phone numbers. This increased availability of

© Springer-Verlag London 2015
G.N. Kenyon and K.C. Sen, *The Perception of Quality*,
DOI 10.1007/978-1-4471-6627-6_16

customer-specific data is causing firms and whole industries to shift their operating philosophies away from traditional goods-based logic to a service-dominant logic.

Traditionally, a firm's competitive advantage would rely upon the exchange of goods; in today's market, a portfolio-based approach is demanded. This portfolio paradigm stresses the application of knowledge and skills on the generation of benefits for consumers. Thus, an understanding of customer perceptions is necessary.

16.1 The Perception of Service Quality

There are three distinct perceptual properties that are present in every service offering: search, experience, and credence. The search properties, though less relevant than with products, include factors that are related to those service attributes that can be determined before the actual service experience. Experience properties include factors that are related to those attributes that can only be evaluated after actually experiencing the service. Credence properties on the other hand incorporate those attributable factors that the consumer often finds difficult to judge even after experiencing the service. It has been argued that most service offerings possess relatively higher experience and credence perceptual properties, while the perceptual property of search tends to be lower, as opposed to manufactured goods that have relatively high perceptual properties in all three areas.

In designing a successful, high-value service system, all three of these perceptual properties must be taken into account. One method for integrating all of these properties into a service design is to introduce the dimension of time in mapping customer interacts with the provider's service offering. The critical point in the time dimension is when the customer first gets a chance to judge the quality of the service, during their initial encounter with the service provider. This could be a phone conversation or an actual visit to the provider's physical location. This is denoted in the literature as the "moment of truth" (MOT).

In all probability, the prospective customer has previously heard of the service provider's offerings via word of mouth or through the provider's promotional activities. This information either establishes credence for the service, or modifies an existing credence. Depending upon the customer's degree of need and his/her level of confidence concerning the service, additional information may be collected through a search process. From the customer's credence and search properties, a set of expectations will be formed. In case these expectations of the service's benefits provide an acceptable level of value, the customer will pursue an experience of the service. It is at the initial point of contact with the service provider that a "moment of truth" begins for the service and the customer.

Given that our expectations are based upon our beliefs, understanding how customers develop their perception over time is important. Thus, each successive service encounter provides another "moment of truth," and the customer's satisfaction will again be based upon the perceived quality of the beneficial attributes versus their expectations. Therefore, the model for understanding consumers' expectations

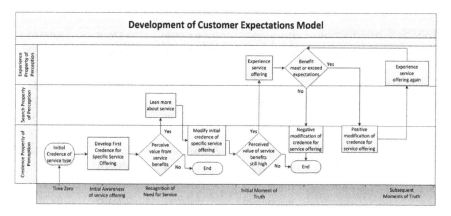

Fig. 16.1   Time-based model of moments of truths

and perceptions must start at the initial point of awareness of the provider's service offering. We therefore divide our model into different stages, based upon the time dimension.

16.2 Moments of Truth

A "moment of truth" occurs every time a customer interacts with a service system. With each of these instances of contact (e.g., advertisements, products, sales force, on-site visits, and service encounters), the customer has an opportunity to evaluate the service provider and the benefits of the service offerings. As shown in Fig. 16.1, these evaluations form the basis for the customer's credence, or opinion, of the service as a whole and the service provider specifically.

Due to the intangibility of services, management cannot directly verify its quality of the attributes defining the service. Often, indirect measurements cannot insure that each "moment of truth" encounter is a high-quality interaction. Only the customer can judge the quality of a service experience, and these judgments can vary greatly from one customer to the next even if the service is executed exactly the same way for each customer. This variance of opinions on the quality of a service by customers is due to the differences between their expectations and motivations at the time of the service.

Customer expectations and their subsequent application in measuring the quality of a service encounter are moderated by the customer's mood and worldview.[1]

---

[1] A person's "worldview" can best be described as a collect of their generalized views of the surrounding world and their perceived place in it. These views are generally based on their relationship to the world, their general beliefs, sociopolitical leanings, moral foundations, and the principles by which they appraise material and spiritual events.

Table 16.1 Credence
properties of service quality

Service Attributes	Capabilities	Convenience	Tangibles
Competence	**0.63360**	0.32610	0.03592
Consistency	**0.57308**	0.29918	0.19362
Courtesy	**0.56383**	0.34885	0.21629
Define	**0.68287**	0.09810	0.35447
Describe	**0.65640**	0.16992	0.19479
Explain	**0.67049**	0.09914	0.25500
Flexibility	**0.50733**	0.42102	0.25159
Readiness	**0.53651**	0.33942	0.39964
Willingness	**0.64940**	0.27504	0.15670
Hours	0.09501	**0.71319**	0.25498
Involvement	0.33664	**0.73105**	0.12159
Layout	0.15028	**0.62345**	0.24459
Location	0.28950	**0.61390**	0.28801
Time	0.36545	**0.65967**	0.18365
Transactional	0.38524	**0.53099**	0.23347
Facility	0.26899	0.31953	**0.52761**
Personnel	0.28488	0.12271	**0.67685**
Representations	0.13783	0.25540	**0.76575**
Tools	0.24259	0.26889	**0.70961**

Managing customer expectations needs a time-phased strategy. There are three
principle phases to this approach: the precontact awareness phase, initial encounter
phase, and subsequent encounter phase.

16.2.1 Precontact Awareness Phase

Initially, the consumer will recognize a need for the benefits of a given type of
service. This need can originate either from a necessity to achieve a specific goal
for which he does not have either the skill set or the time to accomplish, or from a
desire not to do the work himself. Alternatively, it can also originate from a simple
whim. In either respect, either the customer will be aware of the provider (e.g.,
latent credence) that can do the service, or he will seek information about providers that can provide that type of service.

There are several attributes[2] which have been identified that define the quality
of a service offering. These attributes are grouped into three categories of credence
as shown in Table 16.1: the service provider's capabilities, the convenience of the
service, and the tangibles associated with the offering. The capabilities factor is
basically the qualifier for proceeding to either the experiential phase or the search

---

[2] The attributes of service quality are discussed in detail in Chap. 12.

Service Attributes	Capabilities	**Engagement**
Competence	**0.70663**	0.25938
Confidentiality	**0.62437**	0.28997
Consistency	**0.72989**	0.16613
Courtesy	**0.56822**	0.43071
Define	**0.68062**	0.26939
Describe	**0.56837**	0.34398
Explain	**0.57856**	0.35895
Location	**0.58093**	0.32877
Reputation	**0.65947**	0.23377
Time	**0.50645**	0.49471
Transactional	**0.55372**	0.48310
Willingness	**0.63904**	0.38627
Ambiance	0.46665	**0.50905**
Facility	0.36021	**0.58360**
Involvement	0.41042	**0.62600**
Personnel	0.25434	**0.73311**
Readiness	0.45664	**0.54586**
Representations	0.21760	**0.72718**
Tools	0.18514	**0.75424**

Table 16.2 Search properties of service quality

phase of the decision-making process. The convenience and tangible factors are the elements used to determine whether there is a potential enough value associated with the service to justify further considerations.

From information that they receive, their prior experiences, motivations, and other psychological and social influences, customers will form their credence, or beliefs, about a service offering and the service provider. With respect to the quality attributes that define that service, the customer will aggregate their judgments (e.g., good, bad, or indifferent) of these attributes into the previously mentioned categories.

Table 16.1 presents the result of a factor analysis that was carried out for a service-related product. The results show that should the need driving their interest in a given service proved significant enough, the customer will either seek to experience the service (e.g., their credence is strongly positive, or there is little perceived risk associated with the experience), or actively look for additional information concerning the service and possible service providers.

As with credence, the quality attributes of the service are grouped into two categories as shown in Table 16.2: the service provider's capabilities and the engagement factor of the service. The capabilities factor is basically the qualifier for proceeding to experience, while the engagement factor is the order-winning elements. In trying to determine the actual benefits of the service offering, or even which service provider to use in resolving the need, the customer may acquire promotional literature, or search for information on the Internet or from other available sources.

The prospective customer will gather information on the service provider's locations, hours of operation, prices of various items, description of the retail outlet, warranties of services rendered, etc. This information gathering of attributable data is an example of search properties. From this search, the consumer might be motivated to visit the service provider's facility, or Web site, and make a purchase.

During this precontact period, there are three types of decision errors that might occur. The first type of error is the "error of saliency," which involves the consumer not receiving accurate, or complete, information regarding the service and/ or the service provider's. Due to the missing, or inaccurate, information, a gap is created between the service provider's understanding of the customer's expectations and the customer's actual expectations. To prevent this type of error from occurring, information given in promotional literature, or on Web sites, must be accurate and complete about the service's attributes. In addition to these characteristics, the information must address the benefits that consumer really cares about.

The second type of error is the "error of overpromising," which occurs when expectations are raised so high by the information received that the actual delivery of the service cannot live up to them. This overpromising creates a gap between the customer's expectations and the service's actual benefits, which generally leads to overall customer dissatisfaction. One of the principle root causes of overpromising is the espousing of vague promises. To avoid this type of error, the service provider needs to insure that all external communications to customers about the service's attributes and capabilities must be carefully designed. Communications should be clear, unambiguous, and concise. Thus, the motto is "Keep it Simple."

The third error is termed "points of parity," or "points of differences." Points of parity define where the service is on par with other competitive offerings, while points of differences argue those features that set the brand ahead of its competition. The failure point defined by these two objectives occurs when the information provided to the customer is insufficient to stimulate their interest in the service offering. In other words, after an evaluation of all searchable information, the customer is still not motivated to engaging with the service.

Assuming that there is a need for the service, the underlying reason for the customer's lack of interest is often due to the information inadequately conveying the benefits of the service, or its failure to present a convincing case for how the service offering is better than other competitive offerings. Frequently, when the perception property of search suggests that a service offering lags behind other competing offerings on important attributes, or is not offering better value than his/her regular service provider, the prospective consumer's transition to the next step is not likely to occur. Those errors that occur during the awareness stage are components of the search properties of services. These search components suggest the "capabilities" of the service to render benefits and highlight "engagement" factors that may convince the customer to experience the service.

Information concerning the service provider's capabilities should be designed such that they create impressions of the benefits that the service is capable of delivering. This type of information can range from attestations from previous customers, to certificates and awards won by the provider. This presentation of evidence about the service provider's capabilities can create a positive impression for the consumer. Information focused on enticing the customer to engage in an interaction with the service provider typically concerns the convenience of the service, as well as those attributes that provide satisfaction, or elicited excitement about the various transactions involved in service. These attributes include prices, the location, hours of operation, delivery charges, tangibles that accompany the service, and additional or complimentary services. For example, a customer might decide not to go to a restaurant based on the prices posted on an online menu. On the other hand, if there was a special event or entertainment accompanying the meal, the price might represent a significant value. Thus, right at the outset, the service design must carefully think about the information that is included in the search properties as it is instrumental in creating a "customer gap," either positively or negatively.

16.2.2 Customer Contact Phase

If the search information sufficiently engages the customer, he/she could make the first visit to the service provider's facility. This initiates the initial "moment of truth." Should this initial encounter result in a satisfying experience, there may be additional "moments of truths" in the future. If the encounter leads to dissatisfaction, it is essential for the service provider to understand what led to the service failure.

16.2.2.1 Initial Contact

The customer's first interaction with the service provider, either at the store or through a phone call or a visit to the Web site, is often the most important point of contact because it creates the baseline impression of the service system by validating or invalidating credence with actual experience. Should the comparative impression be negative, the customer will most likely balk or renege on the service. This negative comparison will also modify the credence and often be the customer's last impression of the service. Thus, the first contact will be the only point in the theoretical distribution of all "moments of truth." Because the customer usually will not opt to evaluate the service provider again, and may pass on negative reviews to other potential customers, it is important to understand this encounter failed.

During the experiential phase of the customer's perception of a service offering, the attributes of service quality are grouped into three categories as shown in Table 16.3. The first factor grouping includes attributes that measure the service offerings' ability to meet the customer's basic needs for the service (i.e., qualifiers). The second factor grouping includes attributes that often increase the customer's satisfaction with the service (i.e., satisfiers). The third factor groupings of quality attributes are the tangibles that describe or define the service.

Table 16.3 Experience
properties of service quality

Service Attributes	Qualifiers	Satisfiers	Tangibles
Competence	**0.56933**	0.49935	0.01093
Confidentiality	**0.55241**	0.16897	0.29520
Consistency	**0.75429**	0.18827	0.11798
Courtesy	**0.59440**	0.29733	0.17542
Define	**0.64512**	0.27292	0.21071
Explain	**0.59387**	0.24843	0.21387
Financial Safety	**0.65230**	0.11135	0.27176
Follow-Ups	**0.51842**	0.25036	0.30107
Physical Safety	**0.64298**	0.15320	0.19033
Readiness	**0.64022**	0.19381	0.28431
Reputation	**0.57682**	0.08596	0.31306
Willingness	**0.68393**	0.32558	0.11671
Operating Hours	0.14996	**0.67225**	0.24581
Involvement	0.45331	**0.57116**	0.24150
Location	0.21762	**0.69215**	0.17502
Waiting Time	0.39565	**0.59100**	0.11730
Transactional Safety	0.32314	**0.63486**	0.35070
Ambiance	0.22314	0.41670	**0.53340**
Facility Physical Appearance	0.42475	0.11509	**0.55009**
Facility Layout	-0.04409	0.48790	**0.61396**
Personnel	0.21865	0.21881	**0.65951**
Representations	0.29970	0.15787	**0.61498**
Tools	0.27270	0.10006	**0.68872**

All too often, customers may experience a service for the first time, be satisfied with the service offering, and never return, leaving the provider with the impression of a service failure when there was none. This type of failure usually occurs because of circumstances beyond the service provider's control. For example, the customer might move to another residence, where there is no convenient access to the provider's service. Alternatively, if the consumer's negative impression of the service was the result of a design flaw, management needs to understand where the flaw occurred. In the investigation of these types of flaws, management must consider the following questions;

- Was the customer a legitimate member of the target population?
- If the customer was a member of the target population, the following questions are raised.
  - Did the design adequately account for the needs of the population?
  - Did the pre-awareness information accurately describe the service and it benefits?
  - Was the pre-awareness information properly positioned to be noticed by the customer?

- If the customer is not a member of the target population;
  - Should the service design be expanded to include the customer segment that the customer belonged to?

Any failure to gain the consumer's confidence is a failure of design. This failure is mainly because of the design failing to adequately address the service quality attributes that provide customers with the assurances that lead to their satisfaction. A good service design must focus on setting the proper environment for a positive engagement. This means designing the layout and ambiance of the facility to control the customer's mood and provides them with adequate safety both physical and transactional. A proper service system design provides timely information that not only informs the customer of the service offering and its benefits based on the service provider's credentials, capabilities, and expertise, but is also instrumental in influencing the customer's expectations.

All too often, the factors leading consumers to not revisiting a service are the results of either failing to meet the customer's basic needs or providing an unsatisfactory encounter with the service staff. There are two principal factors that may cause a customer not to patronize a service again. First, a lack of credibility with the provider's capabilities claims because of the tangible cues (e.g., artifacts) in the servicescape. For example, dilapidated equipment in a medical center will not inspire confidence in a patient, even if the medical staff is polite and friendly. Second, the servicescape promotes a lack of security causing the consumer to not feel comfortable about the environment of the transaction. Examples of servicescape security failures are as follows: the proximity of the service facility to crime-ridden areas; the type of fellow consumers that frequent the service facility; and the behaviors or appearance of the staff. Thus, other than the inability to meet basic needs and unsatisfactory personal interactions, both tangible cues and environmental reasons are experience qualities that can deter the customer from coming back after the first visit.

A good service system design not only addresses the needs of the customer, but also prepares the customer for the actual service experience. Information that describes the service and its benefits needs to be prepared and made available for the pre-awareness stage, thus setting appropriate expectations. In addition, the actual environment and setting of the service facilities must be staged to moderate the customer's moods. Furthermore, the tools and artifacts used by the service provider must be designed not only to accomplish tasks needed in delivering the service, but also for their ability to re-enforce the information and physical presentation of the service offering. In other words, there needs to be a balance and symmetry between all of the elements constituting the overall service system, such that the customer's perceptions of the service match or exceed his/her expectations of the service as shown in Fig. 16.2.

16.2.2.2 Subsequent Contacts

If the customer has a satisfying initial experience with a service system, he/she might choose to frequent that service again in the future. Even if the customer chooses to revisit the service again, there are no guarantees that he will continue

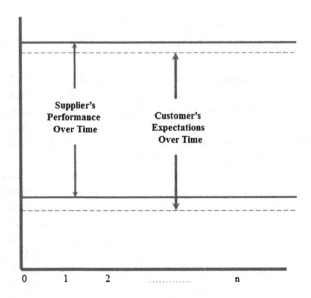

Fig. 16.2 MOT performance meets or exceeds expectations

to revisit the service. This visit thus becomes another test. Unlike with the initial visit, in the customer's mind, the expected benefits of a service are only as good as the last experience. Consistency in the delivery of a service offering, along with the emergence of new points of differences as compared with the competition are major factors in determining the degree to which a customer will repatronize the service.

With each successive encounter, the customer's expectations for the next encounter will be formed. The more consistent the service delivery system is from one encounter to the next, the more the customer will trust the service's reliability. Furthermore, the more reliable the perceived service system, the narrower the expectations of the customer become for the next encounter. To compound this issue, customers will come to expect the same level of service over time and with each successive encounter that does not live up to or exceed previous expectation, the greater the likelihood of defection. Thus, service providers need to incrementally improve the service offering, not only to keep customers loyal, but also to keep the points of differences with competitors small.

Figures 16.3 and 16.4 illustrate the development of these interactions (or "moments of truth") with the service provider over time. As competitive pressures increase in the marketplace, service providers must improve their offerings so as to provide customers with equal or greater benefits. If the pace of a service offering's improvement lags, the marketplace points of differences will occur. Simultaneously, given their credence from prior experiences and additional information from the marketplace, customer expectations will narrow in anticipation of how much better the service benefits will be at the next exchange. As long as the service benefits and performance meet or exceed the customer's expectations, the customer will be satisfied and there will be a strong probability of a future visit. On the other hand, if customer expectations narrow faster than the service's

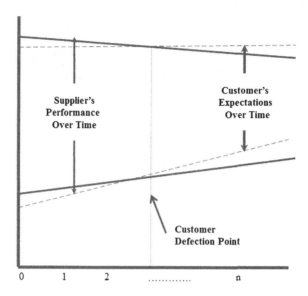

Fig. 16.3 MOT improvement due to consistency lags expectations

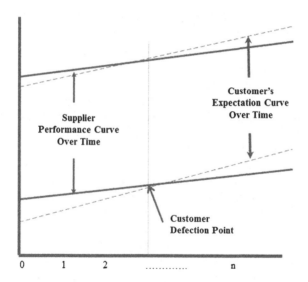

Fig. 16.4 MOT improvement in performance lags expectations

performance can improve, the end result will be customer dissatisfaction and subsequent defection.

Every now and again, a breakthrough improvement will happen. Sometimes, the breakthrough is by design, while other time, it is a singular event that is noted and incorporated into the service system. When these breakthroughs happen they will elicit either a very negative or a very positive response from customers. These departures from the normal delivery design can be called "ultra-moments of truths" (Fig. 16.5).

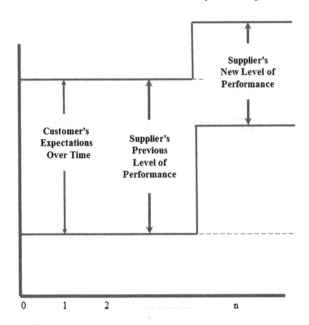

Fig. 16.5 Ultra-MOT

When a negative or traumatic ultra-moment of truth occurs, some part of the delivery process failed. This failure could be either due to an actual failure by the service provider, or due to the customer failing to understand what is occurring or why it is occurring. For example, movie goers might see a sudden interruption in a film's screening because of a malfunction in the movie projector. When a traumatic event occurs, it is important that the service provider makes every effort to attenuate the consumer's negative experience, for example, by giving free tickets for a future movie of their choice to every person in the theater. Alternatively, the traumatic occasion might result because of circumstances which are beyond the provider's control. For example, dinner goers with a reservation at a popular restaurant could be delayed by an unforeseen traffic jam. The restaurant management should make every effort to seat the dining party even though almost every table is taken. The responses to these traumatic "moments of truth" by the provider can often partially obliterate a consumer's bad memories. The provider's extra effort might instill a seed of loyalty in the consumer because of a sense of gratitude for services rendered under difficult circumstances.

Ultra-"moments of truth" can also lead to a pleasant surprise on part of the customer leading to delight (Fig. 16.6). These spikes in performance often arise when a provider provides an unusual benefit to the consumer which the latter did not expect. Often, these can result from purely human endeavors that are dependent on the provider's staff anticipating or remembering a customer's special preferences. For example, hotel management might have a frequent patron's favorite dinner ready for him although he missed his flight and is late for the traditional dinner times at the hotel. Alternatively, the service provider might use technology to reward loyal customers on a particular occasion, either based on some personal data or based on the frequency

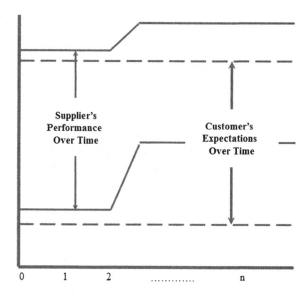

Fig. 16.6 Delightful ultra-MOT

of their visits. These are usually based on the innovative use of information technology. Thus, either human endeavors or technology-based techniques can lead to positive spikes in performance which create pleasant surprises for the consumer. When this occurs, consumer expectations often shift higher. As a result, the provider must constantly aim to convert the upward spike(s) into a regular occurrence, i.e., incorporate the spikes(s) in Scenario 2 into a future band of "moments of truth."

In contrast, if the provider's service had been below the consumer's expectations based on previous expectations, the customer would be disappointed with the service. For example, an ATM machine could be out of cash and disappoint the customer who always frequented it before. Too many traumatic "moments of truth" can lead the customer to stop patronizing the bank. Alternatively, a one off traumatic "moment of truth" can abruptly prevent the customer from revisiting the store again. For example, damage to clothes by a dry cleaning service could stop the customer from revisiting the store. In these cases, the customer moves to defecting from the service provider because of disappointing MOTs. Here, the task of inviting the customer to frequent the provider's service in the future becomes almost impossible, and in all likelihood, the store will lose the customer's future patronage.

The totality of all aspects of the service experience during these moments of truths shapes the customer's perception about the service and will affect whether or not they will continue to patronize the service.

Here, two properties of service encounters play important roles. First, narrow bands of performance indicate consistent experience qualities. This leads to the desired "reliability" quality. Second, the characteristics of the spikes, which can be measured as the ratio of positive to negative spikes, indicate the "responsiveness" and "empathy" quality of services. These two different experience characteristics lead the customer to form strong opinions, either positive or negative, about

the nature of the service. The feelings about the service are particularly strong if
the spikes relate to saliency, i.e., service attributes that are important to the consumer. Over time, if future encounters with the service provider bolster their existing views, customers form strong beliefs about the service provider. These beliefs
about the service are an example of credence qualities. These credence qualities
can be about three different properties of the service provider. For example, the
consumer might feel that the equipment related to the service is modern compared
to the competition. This is an example of a tangible factor.

Second, the consumer might believe that the provider's service locations are
one of the few within the industry that are open around the clock. This is an example of a convenience factor. Third, the consumer might feel confident that her current hairstylist never fails to give her a haircut that makes her look her best. This
is an example of a capability factor. All three factors are examples of credence
qualities that develop over time. It is important that the provider is knowledgeable about both the positive and negative credence qualities that consumers believe
about the service. Also, the provider must attempt to understand the exact nature
of both the search and experience qualities that led to the credence qualities developing over a span of time. Figure 16.7 illustrates the model that shows how consumer perceptions about a service build up over time.

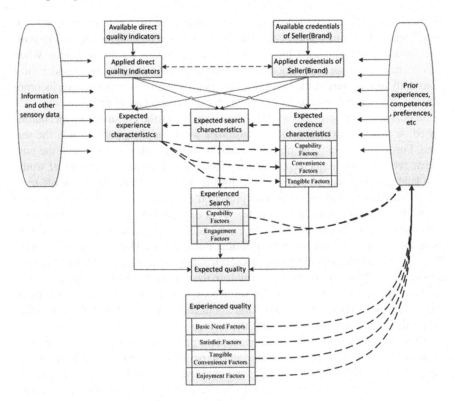

Fig. 16.7 Creating customer perceptions of service quality

16.3 A Framework for Service Design[3]

In order to design a successful new service, or the improvement of an existing service, the provider has to attempt to gain and maintain a sustainable competitive advantage. This is similar to Herzberg's "two-factor" theory. Here, Herzberg distinguishes motivators that give positive satisfaction from factors that do not. Perception properties also have factors that are the basic requirements for consideration of the service and other factors that are associated with deriving satisfaction from the service experience. With credence properties, if the customer does not believe that the service provider is capable of providing a satisfying service experience, he/she will not even worry about whether the service is convenient or if the tangibles are fulfilling. The same logic applies to search properties. If the customer does not find validation of the service provider's capabilities, he/she will not even consider engaging the service in the future. Furthermore, if the customer engages in the service experience and the provider does not meet his/her basic needs, the probability of satisfying him/her is greatly reduced.

Based on the model, described in Sect. 16.3, the design of service systems has four stages. These are as follows: (a) the gathering of information; (b) the translation of information into policies; (c) the implementation of policies; and finally, (d) the continuous monitoring of consumer perceptions and the resultant changes and alterations in policies and practice. While each service sector has its own unique challenges, the framework of the design should avoid particular pitfalls that are likely to creep in. Keeping these "do's" and "don'ts" in mind, we discuss some salient aspects of the design process with reference to the model for each of the four stages.

16.3.1 Getting Information from the Consumer

The process of gathering information from existing and potential consumers should be preceded by careful thought to what the provider needs to know about being successful in the service industry. The first step in this process is to strive to include all the salient attributes that consumers use in choosing a provider. Also, every effort must be made to include all alternative competitive providers that the consumer also considers. The digital age has seen the proliferation of information about various types of inter-competitive forces offering different alternatives for the same basic need. For example, a simple dinner can involve choosing between different restaurants, various food delivery services, microwave choices, etc. Here, the provider must be careful to avoid concentrating only on types of food services that are similar to its own. Thus, it is essential that the "content space" of information which has to be garnered is carefully determined before the actual collection process.

---

[3] Published with Kind Permission from © Taylor and Frances Group, LLC. *Source* Kenyon and Sen [1]. All rights reserved.

The second step in the process is to strive to get the unadulterated consumer view about the actual service. Several types of errors can contaminate the information gathering process. For example, in an effort to lower costs, some providers opt to make the service personnel who interact with the consumer administer the survey. This raises a moral hazard problem, with the actual questionnaire more intent on getting a better rating rather than collecting the consumer's true opinion about the service encounter. Thus, the collection process must be designed to prevent exogenous factors from contaminating the veracity of the information gathered.

The third point to remember in the information gathering process is to focus on the collection of the total distribution of all "moments of truth." For example, consumers are more likely to be willing to report only on ultra-"moments of truth," rather than on service interactions that fall into the band of reliability. The provider must make every effort to garner opinions on both types of interactions. While consumer experiences about ultra-"moments of truth" are likely to provide clues to future remedial action, feedback on "typical" satisfactory interactions is also instrumental in providing information on any possible advantages that the provider needs to continue to maintain over the competition.

It is important to note that reactions to a service can be markedly different among different segments of customers. Thus, one must ensure that all groups are represented in customer surveys. Overall, consumer information has to be analyzed with the aim of discerning the linkage between the three properties of perception—search, experience, and credence—that influence the service selection and patronage process. Thus, the questionnaire or interviews should focus on the attributes that are the underpinnings of the linkages between the three qualities. The collection of information relevant to this should reduce Gap 1, the difference between management knowledge about consumer expectations and actual consumer expectations.

16.3.2 Formulating Policies

Based on the information gathered from consumers, the provider should formulate a set of standards that employees should follow in their interactions with customers. The standards should be set in quantifiable form for each of the typical interactions that the consumer is expected to encounter at the service location. The mapping of consumer expectations to the formulated standards and policies involves a series of studies that measure how actual policies translate into final results that the consumer expects from a provider. Here again, it is important to understand the linkages between the three properties of perception—search, experience, and credence—and focus on the aspects of each that are salient to the customer. Standards that arise from an understanding of the network of linkages between the three properties of perception applicable for a particular service provider will result in a reduction in Gap 2. This is the difference between the management's perceptions of customer expectations and the final design of service design and standards. Basing a provider's policies on competitive standards might not always result in a reduction in Gap 2, as consumer expectations for each provider can be different.

During policy formulation, it is important to distinguish the actual service encounters that involve the customer from those incidences and events that do not. Mapping the service delivery system using a service blueprint format, as represented in Fig. 16.4, can be very useful. The service blueprint is a flowchart of the service delivery process that had been divided into three zones. Each zone defines a level of interaction between the customer and the service delivery system. Zone 1 defines the activities that the customer engages in prior to his/her entry into the service system. Zone 2 defines the service-based activities that the customer is directly involved with. Zone 2 is frequently referred to as "front-end or front-office" operations. Zone 3 defines the service-based activities that are performed without direct customer interaction. Zone 3 is often referred to as "back-end or back-office" operations. An example of these divisions can be seen in bank operations. A customer receives a large sum of money in Zone 1. The customer then enters the bank and informs the teller that he wishes to deposit the money into his account. The teller receives the money, counts it, and then executes a deposit transaction in Zone 2. The transaction is received by a clerk in the bank's back office where the transaction is posted to the customer's account, which is part of Zone 3. Upon receiving notification that the transaction has been correctly posted, the teller hands the customer a receipt of the transaction within Zone 2. Finally, after completing all his transactions, the customer exits the bank in Zone 1. See Fig. 16.8 for a generic diagram of the service blueprint.

An applied example of using the service blueprint in policy formulation is the incidence of accidents for a line of taxi cabs serving a metropolitan area. This might be related to the number of hours that each taxi operator has to drive without a break.

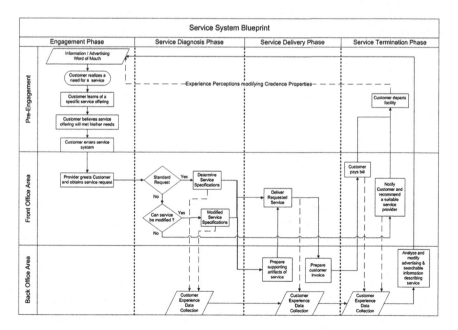

Fig. 16.8   The service system blueprint

A prudent firm would want to improve customer perceptions of the company's safety record by reducing the number of accidents. Thus, by monitoring the operating hours for each driver, the taxi service could work to reduce the incidence of accidents affecting the customer. However, all "front-end" and "back-end" operations may not be related. For example, mistakes happening at the operating theater of a medical clinic could occur solely because of wrong decisions made at the patient's bedside. Thus, the provider must carefully formulate policies for both "front-end" and "back-end" operations and recognize the links between the two types when they do exist. Some of the linkages between the two types of operations might already be required by regulatory agencies. For example, restaurant workers in many counties in the USA are mandated to wear gloves on working on food items, as unhygienic handling of these products is related to health hazards for the customer. Thus, service providers not only have to recognize the major linkages between these two types of operations, but also formulate policies for both kinds so that they do more than merely follow procedures that are set by governmental or civic agencies.

A second dimension in policy formulation is the distinction between a "normal" service encounter and traumatic "moments of truth." Here, two separate types of guidelines come into play. For example, in the case of the movie theater with a malfunctioning projector, there should be guidelines about the number of operational hours after which each projector has to be serviced and inspected. This is an example of a "back-end" operation. However, a second set of guidelines should also be set defining the expected parameters for employee interventions to traumatic "moments of truth." Thus, theater employees will be able to react in a timely and appropriate fashion in the event of a projector breakdown to mitigate customer dissatisfaction. This is an important component of "front-end" operations.

While the first set of guidelines are preventive measures, the second set attempts to reduce the damage caused by unexpected events. Often, the second set of guidelines should be in place to cater to the "perishability" aspect of services. As services cannot be stored, providers have to deal with cases where demand does not match supply. Thus, employees must follow established guidelines to react to situations where an unanticipated consumer demand or a breakdown in supply has the potential to lead to traumatic "moments of truth."

16.3.3 Implementing Policies

Gap 3 refers to situations where the actual performance by the provider's employees does not conform to established service standards. In order to reduce Gap 3, the provider must ensure that employees have the proper training so as to conform to the guidelines. Given the quality of perishability, in many cases, employees should be versatile enough to respond to situations where mismatched supply and demand give rise to the need for possessing a wide array of job skills. In addition, in order to be competitive, the provider must ensure that service employees are supported by the appropriate technology. This aspect is important, as committed employees might not be able to match a competitor's responses to a customer, if they are at a disadvantage because of inferior technological support.

Two components of service design come into play as the policy advances in information technology provide more opportunities for the service provider to satisfy his/her customer. The second is the choice of the mix between centralization and decentralization for the whole chain. Both these dimensions will be discussed in more detail in our analysis of the critical elements of service design.

16.3.4 Monitoring Consumer Perceptions

There are many opportunities for service providers to duplicate and improve upon the unpatented features of a competitor's service offering. Thus, the provider must continuously monitor customer perceptions about both the provider and its competition. Feedback from consumers, the adjustment to standards and policies, and the monitoring of performance must be managed as a continuous process. One way to ensure that the feedback loop is working properly is to periodically have new customers evaluate the quality of the service with regard to experiences with other providers.

During the monitoring stage, the provider should identify the underlying reasons behind the ultra-"moments of truth" that result in customer delight. For example, diners at a particular restaurant location within a service chain might like the extra special attention given to them by the wait staff. The provider should try to ascertain the practice followed by the restaurant management which led to this customer experience. For example, the provider could find that management at that location made two members of the wait staff share the responsibilities for each table. Thus, if a waiter/waitress is unavailable, a patron has the option of asking another wait staff for service. Given this scenario, the provider should attempt to implement a similar policy at other locations within the chain.

Overall, the service provider must strive to learn what promises (during the search stage) and interactions (during the experience stage) make consumers hold certain beliefs about a service (during the credence stage). A deeper understanding of this linkage will make the provider more knowledgeable about the competitive forces at play and the requisite course of actions that are required to succeed in his/her service sector. The complete service design incorporating all four stages is presented in Fig. 16.9.

16.3.4.1 Striving for the "Sweet Spot" in Service Design

While the different stages of service design have been described earlier, it is important to consider the various dimensions that come into play for the actual design. A better understanding of these underlying factors provides important insights for the service provider. The first dimension relates to the knowledge component. Advances in information technology, specifically in the gathering, organization

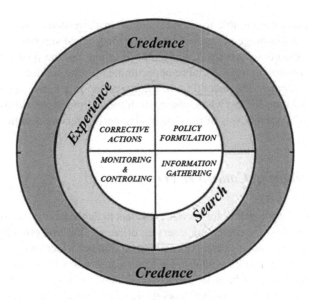

Fig. 16.9 Service system design versus consumer perceptions

and analysis of data, provide opportunities for a more scientific basis of bridging the gap between the service provider and the customer. However, it is important to note that certain types of knowledge cannot easily be transmitted or stored. Thus, "on-the-spot" observations over many years can give astute store managers unique insights into both "back-end" and "front-end" operations. This is often based on detailed knowledge of work staff, equipment, and individual customers served and often cannot be easily codified. Thus, the provider has to recognize the limitations of information technology that prevents service design from becoming an exact science.

Given the potential for the existence of "tacit" knowledge in downstream operations, an overzealous adherence to standards established by a central authority might sometimes result in employees not responding to the specialized needs of customers or operating at subpar efficiency. These incidents will negatively affect consumer perceptions of the service provider on the responsiveness and empathy dimensions. In order to avoid these situations, a second dimension for service design has to be introduced. While the first dimension relates to the efficient use of information on both "front-end" and "back-end" operations, based on the best available technical tools, the second relates to areas where there are limits to "codified" knowledge. Here, in contrast to depending on a centralized system of "codified knowledge," a better alternative might be to allow the "tacit knowledge" of the downstream, "on-the-spot" manager, to play a part. Thus, a change in the organizational format from a centralized to a decentralized structure might be the corrective action needed in this situation.

The dimension of centralized/decentralized organizational format becomes an important second factor. The change in format becomes particularly important in unusual situations where established procedures are not appropriate. For example, a hardware store chain might have a policy of selling on strictly cash or credit. However, after a hurricane, the chain might allow store managers in affected areas to go against established policy and accept checks. This is because credit card transactions become impossible during long periods of communications breakdown. While this temporary change in policy might give rise to some bad checks, the goodwill gained within the community is likely to outweigh the additional financial costs involved. Thus, the second component touches on the issue of a shift in the centralization/decentralization mix in particular situations.

A possible alternative to centralized control is to allow for selective decentralization in cases where the bundle of codified knowledge normally available within the chain is suddenly not effective in providing value to the customer. Notwithstanding the temporary shift within the centralization/decentralization dimension, the chain should continue to use available information technology to build a database of incidents related to extraordinary situations, together with an inventory of effective actions that were taken as a response. By consulting this database in the future, the service provider will reduce the dependence on the "gut feeling" of downstream managers. For example, by following the delivery/reception records of all consumers, post offices, and their own warehouses, a DVD mailing service can make an informed guess about the source of a problem if an individual customer has returned a DVD that has not been received. However, if the provider decides to have a more decentralized structure, he/she must also decide on the exact strata of employees who should be empowered to make decisions in a given situations. Finally, each of these decisions should be analyzed after the event, to ensure that individual managers, perhaps inadvertently, have not indulged in unequal treatment of the service provider's customers. In sum, the effective use of information technology should help build a base of knowledge which incorporates most situations facing the service provider into a scientific paradigm.

Over time, the service provider can design operational policies as a near science. Thus, over the long run, the service is converted from its dependence on individual, uncorrelated decisions based on the "tacit" knowledge of individual managers into a science based on "codified" knowledge, thus improving the consistency of service delivery. Nevertheless, as unanticipated situations arise in future service encounters, the provider must always be prepared to move the point of centralization or decentralization in situations where previously designed policies are no longer applicable. Thus, an adherence to a fixed policy where all decisions are based on a centralized bank of "codified" knowledge is perhaps an unachievable goal for the provider.

There is a wide variation in customer sensitivity to service times. The variance between different customer segments can be particularly significant across national outlets within the same service chain. Thus, the service chain must avoid adopting a "boiler plate" approach for all of its outlets. The difference between generalization and discrimination across different outlets or situations is the third dimension in the service design.

Careful analysis of information collected from consumers can help the provider get a better understanding of nuances in customer preferences and expectations across different locations and times. Generalized solutions are applicable for uniform preferences. However, the provider must not hesitate to provide alternatives when customer preferences are markedly different. For example, a nationwide coffee chain might offer alternative flavors in areas where there are differences in taste. The change between generalization and discrimination is an important one to consider for the service provider. Sometimes, a shift toward discrimination is accompanied by a greater dependence on a decentralized structure that allows downstream managers greater leeway to cater to variances in local tastes.

The fourth and perhaps most important dimension within service design is the role of value creation. It has been noted the potential for services to co-create value for the benefit of all parties involved in an interaction. Unfortunately, the provider can make the correct decisions on all of the first three dimensions in order to create value for everyone involved in the distribution chain, except for the consumer. In contrast to this approach, a bank might decide not to enter the residential mortgage business in a region, as it might feel that current terms and conditions reduce the chances of individual customers paying off their loans and owning their homes during their lifetimes. As the consumer is an essential part of the service interaction, the creation of value for both the customer and the provider is a critical component that should be considered in service design.

The four dimensions to be considered in service design relate to the nuances in the nature of knowledge involved in a service transaction, the degree of centralization/decentralization allowed within the service provider's organization, the mix of generalization/discrimination that the provider allows throughout the chain, and finally the role of value for all parties involved in the transaction (particularly the customer). A movement toward the correct mix on each of these dimensions will ultimately hit the "sweet spot" of service design, i.e., the point at which the provider delivers value to the customer in the most efficient manner possible. However, it is important to note that the "sweet spot" area could vary between different customers. For example, there are subtle differences in attitudes in mobile banking between potential and repeat customers. Moreover, the "sweet spot" can change over time in response to changes in the competitive landscape and customer preferences. Finally, advances in information technology are likely to change the optimal point within the centralization/decentralization continuum. Thus, the provider must always strive to discern the "sweet spot" by continuously examining the point on each dimension where it should position itself. Figure 16.10 illustrates the concept of the "sweet spot" in service design.

The suggested framework for the service design is a prescriptive plan for building a system which seeks to continuously improve a provider's service quality. We identify four dimensions on which the service provider must carefully decide to position itself in order to reach the "sweet spot" for providing the best value to the customer. Future researchers can examine the differential impact of various elements of the model in context of the final design across different service sectors. In some sectors, such as health services, the tangible elements might be critical in determining quality, as part of the servicescape. In other sectors, such as financial and legal services, personal interaction might be the most important element within

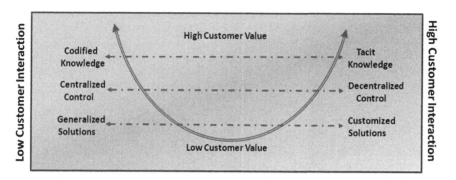

Fig. 16.10 Customer value creation model

the final design. Overall, the suggested design framework contains procedural details that can be applied to almost all sectors. Practitioners can therefore use the suggested design framework in different situations. The important element in the design process is to garner accurate consumer feedback about the service and use this to build a system that maintains a sustainable competitive edge for the provider. This will result in the buildup of positive credence qualities based on the search and experience qualities that have been carefully formulated at earlier stages.

Reference

1. Kenyon, G. N., & Sen, K. C. (2012). Customer perceptions, dimensions of service quality, and service systems. In A. Juan, T. Daradoumis, M. Roca, S. Grasman, & J. Faulin (Eds.), *Decision making in service industries: A practical approach.* Boca Raton, FL: Taylor & Frances Group, CRC Press.

Chapter 17
Integrating All the Components of Quality

The roots of quality go back to the role the product/service plays in providing value to the customer. In this book, we have attempted to examine the various facets of quality, right from its inception and design, through its transformation from raw material components to the final product and the final delivery to the customer. In concluding our thoughts, we revisit some of these topics in order to integrate the various components of quality. We will also examine the possible pitfalls on the route to producing quality and identify the core requirements to producing a good quality product.

17.1 The Product Offering as a Bundle of Attributes

A product/service offering can be viewed as a bundle of the various attributes it possesses. For example, a TV set-top box is an entrant into the world of video/audio which aims to give consumers options for watching programs without using a cable or satellite connection. Thus, it depends on the internet to download shows, news reports, etc., from various content providers. At the bare minimum, each set-top box must have options for connecting to the internet, a HDMI connection to bring a digital signal to the consumer's television as well as a power connection. However, even at this rudimentary level, each supplier has a chance to gain or lose an advantage over its competitors. For example, the supplied power connector that provides electricity to the box might be so large that plugging it into the power strip could interfere with using adjacent power outlets on the strip. This will be a significant disadvantage for consumers who use this particular set-top box. One step above the basic attributes required for all set-top boxes are the additional attributes that each set-top box possesses. Each of these attributes has the potential to provide a significant benefit to the target market. Some of these attributes will be directly related to product technology. For example, one set-top box can have

G.N. Kenyon and K.C. Sen, *The Perception of Quality*,
DOI 10.1007/978-1-4471-6627-6_17

the built-in software technology that has the capability for providing close caption subtitles to all movies streaming from the various channels it carries. Alternatively, a particular set-top box might have its own option for digital audio output such as an optical audio socket. This gives consumers who do not have television sets with this audio output feature to directly connect their set-top box to their home theater audio receivers and hear a better sound output. However, an important attribute that is not directly related to technical design could simply be the number of providers each set-top box carries. A set-top box that has a number of well-designed tangible features could pale in comparison with another with a slightly worse design but with a significantly higher number of popular program providers. Thus, in addition, to the tangible design, the bundle of attributes of each product can be related to strategic partnerships for content that each set-top provider has made. For almost any product, these strategic partnerships with other companies have an important role in achieving a competitive advantage in the market place.

The comparison of advantages/disadvantages arising from the attributes provided by each TV set-top box is related to the concept of points of parity (POPs) and points of difference (PODs). Here, manufacturers must strive to match the competition on important attributes POPs, while offering attribute(s) that are not available in the market place PODs. In a dynamic environment, where each manufacturer strives to get a competitive advantage, today's POD is an important attribute for other manufacturers to copy or imitate, and might not provide a permanent advantage to the original innovator.

Alternatively, a new innovation that is a POD might detract from the way the product interacts with an existing setup. Let us consider the case of a set-top box that has a remote control that is based on Bluetooth technology. This feature offers a significant advantage because direct access to the set-top box is not required for the remote for control. However, some users of home theater systems who are wedded to old universal remote controls (URCs) that control all their visual/audio devices might balk at adopting the new set-top box because their existing URC will not be able to control it. Thus, a potential POD might not have the impact it could have because the new feature does not dovetail with the viewing habits of some members of the target market. At the same time, bold innovators are often prepared to offer advantages which are difficult to copy or disregard. Firms offering these types of POPs need to realize that they will not be an important attribute in the future. An example of the latter is Apple's introduction of the IMac in 1998. While the IMac had a USB port, it did not have a floppy disk drive. The latter was almost a given in all computers shipping at that time and must have been an essential POP. However, Apple guessed correctly and slowly but surely the floppy disk drive has lost its popularity. Thus, manufacturers must offer products whose designs offer innovative new features and also drop features which, primarily because of alternative technological options, are not going to be essential POPs in the future.

The product's overall performance and its perceived value to the consumer are closely related to the availability of the appropriate attributes in the original product. However, consumer perceptions are more than just the sum of product attributes. Thus, all product attributes must be put together to provide not only overall convenience and ease of use to the customer, but also present a pleasing aesthetic appearance. In sum, all the attributes of a TV set-top box must do more than just provide a consistent option over time for the consumer to receive entertainment and news through the internet. The product must also look pleasing to the consumer in the room in which it is placed. Thus, the practical and aesthetic sides of the bundle of attributes must both exist in the final product offered to the consumer.

17.2 Ascertaining Consumer Views

The success of a product is related to how closely it meets consumer expectations about the value that it provides. A misconception often made is that quality is unrelated to how the consumer perceives the product. For example, the manufacturer can produce a product that is made of the best materials, processed according to stringent standards, but for some reason, does not meet customer expectations. This is often the case with automobiles, where consumers might like the actual car itself, but is let down by the buying experience at the dealership. Thus, in order to be successful, quality must be "relevant" (i.e., it should meet the target market's expectation on all the dimensions that matter to the customer). In order to know what these dimensions are and how the product/service offering is performing, the company must continuously monitor customer expectations and relate them to the customer's background and motivations.

Ascertaining views on what dimensions of the consumption experience are important and how well the product/service is meeting them can be done both by communications and observation methods. Communication methods cover questionnaires and surveys which ask existing or potential customers about their opinions about the company's own as well as competitive products about their performance. Survey questions also cover the varying degree of importance which the consumers place on different dimensions. These might change over time and across different customer segments. For example, an affluent retirement community might value prompt and accurate delivery of groceries rather than their price.

Observation methods focus on observing how the consumer is actually using the product. For example, a satellite provider could focus on finding what pay-per-view movies or events different households watch at different price levels. With most transactions being recorded on the basis of UPC or RFID codes, retailers and manufacturers have the ability to observe what combination of marketing instruments, including price, package design, and shelf placement, is being used in each instant a transaction takes place. From the combination of these two methods, companies have the potential to find out not only whether their products are meeting sales expectations, but also explore the underlying causes behind the revenue numbers, by questioning consumers on the reasons behind their likes and dislikes. This should provide a blueprint for the design of new products based on consumer weightage of the importance of attributes as well as feedback on how existing products are meeting expectations.

However, the potential for using communication and observation methods for monitoring existing items and giving inputs for new products might be wasted if the company does not pay attention to several other factors. First, the use of communication methods must not be biased for existing personnel to monitor consumer comments about their performance. This can easily happen if the responsibility of eliciting consumer comments about product performance is given to the salespeople who last interacted with the customer. Even if no explicit pressure is brought to bear on the customer during the administration of the survey, customers might feel uncomfortable in giving their honest opinions about the buying experience when the questionnaire is administered by an individual(s) who played a part in the transaction. Thus, it is important that within a short period of time, the survey is administered by a third party.

Second, a company can fall into the trap of using mandated communications and observation methods by rote. In these cases, while accurate data are collected, no use is made of it to improve existing product performance or provide inputs for new product design. Thus, the collection of data must be inextricably linked to analyst departments who have the responsibility to provide inputs to top management not only on present product performance but also on suggestions for innovation, etc.

Third, too often, management might get bogged down in tracking averages. Thus, if overall product performance meets the minimum acceptable standards, management might feel satisfied and think that nothing more needs to be done, following the philosophy of "how can you improve on perfection?" However, in many cases, if management got over its obsession with averages, they will realize that the data can lend itself to identifying both positive and negative outliers. In both cases, further analysis of the outliers can provide important insights into improving overall performance. For example, in a restaurant chain, the excellent performance of a few outlets might prod management to look at differences in management practices followed by its local management team. These might be adopted by the restaurant company to improve the overall chain's performance. In the opposite type of cases, management should take steps to find out the underlying causes of the poor performance of the negative outliers and rectify them immediately. The identification of these outliers can therefore be a useful step in improving customer perceptions of quality throughout the chain. Depending on the extent of the details of the collected data, the potential to improve overall performance in all spheres is almost unlimited, provided management is prepared to look beyond just the averages.

17.3 Bringing Ideas to Fruition

In many instances, after management gets an idea for a new product or service, they might describe the concept or present a rudimentary prototype to a segment of the target market and meet with their approval. However, after getting past this step, the company might find that their biggest challenge might be to convert the idea to a product/service that is available to the target market on a mass scale. For instance, the delivered product might not conform to the promised specifications on a consistent basis. This might shatter customer perceptions about the product which can cause irreparable damage on the long run.

In an earlier chapter on transformation processes, various improvement methodologies were described. After ensuring that the selected suppliers for raw materials and components are capable of meeting the requisite standards for delivery in terms of product specifications and schedules on a consistent basis, management must select the transformational methodology which is appropriate to its goals and capabilities. This should dovetail not only with the company's capabilities but also with the demands on quality imposed by market conditions.

In the case of service industries, many companies consider using a different organizational format such as franchising to ensure that quality standards are consistently made throughout the chain, and at the same time, expansion goals in terms of outlet growth which are critical in gaining a competitive edge are met. Thus, for both product and service offerings, companies must have a strategic plan in place for delivering the quality that was earlier promised in the prototypes or presentation of the product concept.

It is important to note that each company might adopt a different transformational methodology based on the unique challenges and constraints they face. Thus, it is not surprising that different firms have vastly different mixes of franchised and company-owned outlets. This is because they have each taken different approaches to the quest to maintain chain-wide consistency and also achieve outlet growth.

17.4 Quality as a Dynamic Concept

In the industrial world, there are many examples of products going into obsolescence: VHS tapes, typewriters, dial up modems, etc. In some cases, these products were surpassed because other options carried out the same function better. In other cases, technology advances improved the economics, making a different solution superior. In still other cases, the product evolved, adding new functions. In the case of the smart phone, the original purpose of making and receiving phone calls has been complemented by the gamut of functions that cover text messaging, web surfing, GPS, and other activities governed by different applications (apps). Thus, for every product, the manufacturer must constantly monitor the importance of each attribute to ensure that customer expectations about performance are met.

This is also applicable to service industries where customer concerns about identity theft might make the importance of secure transactions for retailers increase in comparison to other attributes. Previous encryption and identity protection method adopted by a particular retailer might not be adequate, and therefore, better security methods will be necessary. Thus, the minimum quality requirement for a particular attribute has increased, and an important component of the buying experience has to be improved quickly. Customer expectations about transactions will necessitate a redesign of the product. In almost all cases, the minimum quality requirement for each product attribute must be constantly monitored alone with that of competitive offerings.

In highly competitive fields, an innovative concept by a competitor, such as touch-sensitive heart rate monitor on a smartphone, might suddenly put pressure on other manufacturers to add an attribute which was not part of their product design. Similarly, as mentioned earlier in the case of computers, some attributes such as floppy disk drives might wane in importance.

There are several factors, such as technological innovation, competitive action, governmental legislation, and other environmental factors that can make consumer change their expectations about product/service expectations. Companies should, therefore, constantly monitor these expectations in a proactive manner so that they are not blindsided by some external change they did not anticipate. Each of them must realize that quality as a concept is multi-dimensional in terms of the changing mix of its construct as well as the minimum levels acceptable in the market place. Successful companies will be the ones who anticipate these changes by incorporating practices that maintain requisite quality levels that provide value to the customer in a cost-effective manner.

The bottom line is that quality is a construct that is unique to the person experiencing it (i.e., the customer). Thus, companies must understand who their customers are and how they perceive quality. The more customers perceive the quality of an item the same way, the more static the definition of that aspect of quality. Compared to dynamic quality definitions, static quality definitions lend themselves better to standardized specifications.

17.5 The Price–Quality Relationship

In many cases, consumers can buy the same type of product at vastly different prices. For example, a watch might sell at a discount store for about ten dollars, whereas a luxury watch could be selling for almost half a million dollars. An easy trap to fall into is to label the cheap and expensive watches as low- and high-quality products, respectively. However, even at the low price range, if the watch lives up to expectations, it could be said to have delivered quality to the customer. The expensive, or the luxury, watch offers dimensions of quality, that the cheaper watch does not offer. These include craftsmanship, the distinctiveness of a luxury brand name, and the prestige associated with it. This essentially implies that although both might belong to the same product category, each type of watch is catering to different target markets and meeting different types of needs and expectations. The same differences apply to an average car used by the typical household for running chores and an ultra-luxury car.

While the different dimensions of quality separate luxury brands from ones that are more often used by the buying public, previous research has found a low relationship between price and quality for brands within frequently purchase product categories. Significant positive relationships exist for more expensive, non-frequently purchased items. Thus, for a wide array of items, there is no evidence that more expensive items have higher quality. This implies that it is possible to achieve the quality requirements of consumers for many products we buy every day in a cost-efficient manner. Here, the role of external communications is important. For more expensive, non-frequently purchased products, the role of promotions is to emphasize the extra product features and advantages that an increase in price could bring to the customer. For the more frequently purchased products, companies would do well to concentrate on improving quality on all the relevant dimensions in a cost-effective manner and use promotion to emphasize their low-price advantage over more expensive brands.

Just like beauty, quality is in the eye of the beholder. By understanding the dimensions by which customers perceive and measure quality, and the way they perceive changes in those dimensions, you can maximize the competitiveness of your company by increasing the satisfaction and loyalty of your customers.

Bibliography

1. Akao, Y. (2004). *Quality functional deployment: Integrating customer requirements into product design.* New York, NY: Productivity Press.
2. Akao, Y. (1997). *QFD: Past, Present, and Future. International Symposium on QFD, 97,* 1–12.
3. Aktas, E., Cicek, I., & Kiyak, M. (2011). The effect of organizational culture on organizational efficiency: The moderating role of organizational environment and CEO values. *Procedia Social and Behavioral Sciences, 24,* 1560–1573.
4. Amabile, T. M. (1997). Motivating creativity in organizations: On doing what you love and loving what you do. *California Management Review, 40* (1), 29–58.
5. Amabile, T. M., & Conti, R. (1999). Changes in the work environment for creativity during downsizing. *Academy of Management Journal, 42*(6), 630–640.
6. Anadalingam, G., & Kulatilaka, N. (1987). Decomposing production efficiency into technical, allocative, and structural components. *Journal of the Royal Statistical Society, Series A, 150*(2), 143–151.
7. Andersen, E. S., & Philipsen, K. (1998, January). The evolution of credence goods in customer markets: Exchanging "Pigs in Pokes", *DRUID Winter Seminar,* Middelfart, 8–10.
8. Anderson, A., & Media, D. (2013). *The impact of inventory management on customer satisfaction.* Houston Chronicle. http://smallbusiness.chron.com/impact-inventory-managementcustomer-satisfaction-22798.html
9. Anderson, D. (1999). *Synchronized supply chains: The new frontier.* Ascet, (Vol. 1). Stanford University: Accenture Hau Lee.
10. Andrews, J. D., & Moss, T. R. (1993). *Reliability and risk assessment.* Essex: Longman Scientific and Technical.
11. Angel, L. C., & Chandra, M. J. (2001). Performance implications of investments in continuous quality improvement. *International Journal of Operations and Production Management, 21*(1), 108–125.
12. Ashkenas, R. (1999, Jan/Feb). Breaking down barriers. *Industrial Management, 41*(1), 24–31.
13. Bagsarian, T. (2001, March). E-commerce: The growth of private company exchanges. *Iron Age New Steel, 17*(3), 22–23.
14. Baines, T. (2004). An integrated process for forming manufacturing technology acquisition decisions. *International Journal of Operations and Production Management, 24*(5), 447–467.
15. Barney, J. B. (2002). *Gaining and sustaining competitive advantage* (2nd ed.). Reading: Addison-Wesley.

© Springer-Verlag London 2015
G.N. Kenyon and K.C. Sen, *The Perception of Quality,*
DOI 10.1007/978-1-4471-6627-6

16. Bartneck, C. (2009). *Using the metaphysics of quality to define design science: Proceedings of the 4th International Conference on Design Science Research in Information Systems and Technology.*

17. Beatty, S. E., Kahle, L. R., Homer, P., & Misra, S. (1985). Alternative measurement approaches to consumer values: The list of values and the rokeach value survey. *Psychology and Marketing, 2*, 181–200.

18. Becker, B. W., & Connor, P. E. (1981). Personal values of the heavy users of mass media. *Journal of Advertising Research, 21*, 37–43.

19. Belohlav, J. A. (1993). Developing the quality organization. *Quality Progress, 26*(10), 119–122.

20. Benson, P. G., Saraph, J. V., & Schroeder, R. G. (1991). The effects of organizational context on quality management: An empirical investigation. *Management Science, 37*(9), 1107–1124.

21. Bergman, B., & Klefsjo, B. (1994). *Quality from customer needs to customer satisfaction.* London, UK: McGraw-Hill.

22. Bernstein, D. A. (2010). *Essentials of psychology* (3rd ed.). Independence: Cengage Learning.

23. Besanko, D., Dranove, D., & Shankley, M. (2000). *Economics of strategy* (2nd ed.). New York, NY: Wiley.

24. Bolumole, D. J. (2001). The supply chain role of third party logistics providers. *International Journal of Logistics Management, 12*(2), 87–102.

25. Bowbrick, P. (1992). *The economics of quality, grades and brands.* London, UK: Routledge.

26. Bowersox, D. J., Closs, D. J., & Cooper, M. B. (2010). *Supply chain logistics management.* Boston: McGraw-Hill.

27. Bowersox, D. J., Closs, D. J., & Stank, T. P. (2000). Ten mega-trends that will revolutionize supply chain logistics. *Journal of Business Logistics, 21*(2), 1–16.

28. Bowersox, D., & Cooper, M. B. (1992). *Strategic marketing channel management.* San Francisco: Jossey-Bass.

29. Boyson, S., Corsi, T., Dresner, M., & Rabinovich, E. (1999). Managing effective third party logistics partnerships: What does it take? *Journal of Business Logistics, 20*(1), 73–100.

30. Bradley, P. (2008). Constancy, categories and bayes: A new approach to representational theories of color constancy. *Philosophical Psychology, 21*(3), 601–627.

31. Cameron, K. S., & Quinn, R. E. (1999). *Diagnosing and changing organizational culture: Based upon competing values framework.* Reading: Addison-Westey.

32. Cameron, K., & Sine, W. (1999). A framework for organizational quality culture. *Quality Management Journal, 6*(4), 7–25.

33. Campanella, J. (1999). *Principles of quality costs: Principles, implementation, and use* (3rd ed.). Milwaukee: ASQ Quality Press.

34. Carman, J. M. (1978). Values consumption patterns: A closed-loop. In H. K. Hunt (Ed.), *Advances in consumer research.* (Vol. 5). Ann Arbor: Association for Consumer Research.

35. Carr, L. P., & Tyson, T. (1992). Planning quality cost expenditures. *Management Accounting, 77*(2), 52–56.

36. Carter, P., Carter, J., Monczka, R., Slight, T., & Swan, A. (2000). The future of purchasing and supply: A ten-year forecast. *Journal of Supply Chain Management, 36*(1), 14–26.

37. Carter, J. R., Ferrin, B. G., & Carter, C. R. (1995). The effect of less-than-truckload rates on the purchase order lot size decision. *Transportation Journal, 34*(3), 35–44.

38. Carter, J. R., & Price, P. M. (1993). *Integrated materials management.* London, UK: Pitman.

39. Chandra, C., & Kumar, S. (2000). Supply chain management in theory and practice: A passing fad or a fundamental change. *Industrial Management and Data Systems, 100*(3), 100–113.

40. Chandrashekar, A., & Schary, P. B. (1999). Toward the virtual supply chain: The convergence of IT and organization. *International Journal of Logistics Management, 10*(2), 27–39.
41. Chen, I. J., & Paulraj, A. (2004). Toward a theory of supply chain management: The constructs and measurements. *Journal of Operations Management, 22*, 119–150.
42. Chiu, L. H. (1972). A cross-cultural comparison of cognitive styles in Chinese and American children. *International Journal of Psychology, 7*, 235–242.
43. Choi, T. Y., & Ebock, K. (1998). The TQM paradox: Relations among TQM practices, plant performance, and customer satisfaction. *Journal of Operations Management, 17*(1), 59–75.
44. Cooper, M. C., Lambert, D. M., & Pagh, J. D. (1997). Supply chain management: More than new name for logistics. *International Journal of Logistics Management, 8*(1), 1–14.
45. Coren, S., & Girgus, J. S. (1980). Principles of perceptual organization and spatial distortion: The gestalt illusions. *Journal of Experimental Psychology: Human Perception and Performance, 6*(3), 404–412.
46. Coyle, M. C., Bardi, E. J., & Langley, C. J. (1996). *The Management of Business Logistics* (6th ed.). Saint Paul: West Publishing.
47. Crosby, P. B. (1979). *Quality is free*. New York, NY: McGraw-Hill.
48. Crosby, P. B. (1996). *Quality is still free*. New York, NY: McGraw-Hill.
49. Croxton, K. L., Garcia-Dastugue, S. J., Lambert, D. M., & Rodgers, D. S. (2001). The supply chain management processes. *International Journal of Logistics Management, 12*(2), 13–36.
50. Daft, R. L., & Weick, K. E. (1984). Toward a model of organizations as interpretation systems. *Academy of Management Review, 9*, 284–295.
51. Dale, B. G., Lascelles, D. M., & Lloyd, A. (1994). Supply chain management and development. In B. G. Dale (Ed.), *Managing quality*. Upper Saddle River, NJ: Prentice-Hall Inc.
52. Darby, M. R., & Karni, E. (1973). Free competition and the optimal amount of fraud. *Journal of Law and Economics, 16*, 67–88.
53. Daugherty, P. J., Autry, C. W., & Ellinger, A. E. (2001). Reverse logistics: The relationship between resource commitment and program performance. *Journal of Business Logistics, 22*(1), 107–123.
54. Dean, J. W., & Bowen, D. E. (1994). Management theory and total quality: Improving research and practice through theory development. *Academy of Management Review, 19*(3), 392–418.
55. Deming, W. E. (1986). *Out of the crisis*. Cambridge: Massachusetts Institute of Technology, Center for Advanced Engineering Studies.
56. Demsetz, H. (1991). The theory of the firm revisited. In O. E. Williamson & S. G. Winter (Eds.), *The nature of the firm* (pp. 159–178). New York: Oxford University Press.
57. Deshpande, R., Farley, J. U., & Webster, F. E. (1993). Corporate culture, customer orientation, and innovativeness in Japanese firms: A quadric analysis. *Journal of Marketing, 57*(1), 23–27.
58. Diallo, A., Khan, Z. U., & Vail, C. F. (1995). Cost of quality in the new manufacturing environment. *Management Accounting* (August), 20–25.
59. Diamond, P. A., & Mirrlees, J. A. (1971). Optimal taxation and public production I: Production efficiency. *The American Economic Review, 61*(1), 8–27.
60. Dobbs, W., & Monroe, K. B. (1985). The effects of brand and price information on subjective product evaluations. *Advances in Consumer Research, 13*, 85–90.
61. Downey, C., Greenberg, D., & Kapur, V. (2003, Sep/Oct). Reorienting R&D for a horizontal future. *Research Technology Management, 46*(5), 22–28.
62. Duncker, K. (1941). On pleasure, emotion, and striving. *Philosophical and Phenomenological Research, 1*, 391–430.
63. D'Avanzo, R., von Lewinski, H., & Van Wassenhove, L. (2003). The link between supply chain and financial performance. *Supply Chain Management Review, 7*(6), 40–47.
64. Ellram, L., LaLonde, B., & Weber, M. (1989). Retail logistics. *International Journal of Physical Distribution and Logistics Management, 19*(12), 29–39.

65. Emery, F. E. & Trist. E. L. (1960). Socio-technical systems. In C. W. Churchman & M. Verhurst (Eds.), *Management science, models and technique*, (Vol. 2, pp. 83–97). London, UK: Pergamon Press.

66. Evans, J. R., & Lindsey, W. M. (2008). *Managing for quality and performance excellence* (9th ed.). Mason: South-Western Cengage Learning.

67. Farris, M. T. (1997). Evolution of academic concerns with transportation and logistics. *Transportation Journal, 37*(1), 42–51.

68. Fearne, A. (1998). The evolution of partnerships in the meat supply chain: Insights from the British beef industry. *Supply Chain Management, 3*(4), 214–231.

69. Feigenbaum, A. V. (1983). *Total quality control*. New York, NY: McGraw-Hill.

70. Finbarr, L. (2006). Defining high value manufacturing, White Paper, Cambridge University–The Institute for Manufacturing.

71. Fine, C. H. (1986). Quality improvement and learning in productive systems. *Management Science, 32*(10), 1301–1316.

72. Fisher, M. L. (1997). What is the right supply chain for your product? *Harvard Business Review, 75*(2), 105–116.

73. Flaherty, M. T. (1996). *Global operations management*. New York: McGraw-Hill.

74. Flynn, B. B., Schroeder, R. G., & Sakakibara, S. (1995). The impact of quality management practices on performance and competitive advantage. *Decision Sciences, 26*(5), 659–691.

75. Ford, D. (1990). *Understanding business markets*. London, UK: Academic Press.

76. Forrester, J. (1961). *Industrial dynamics*. New York, NY: Wiley.

77. Gaither, N., & Frazier, G. (2002). *Operations management* (9th ed.). Mason, OH: South-Western.

78. Galbraith, J. R. (1977). *Organization design*. Reading: Addison-Wesley.

79. Gale, B. T. (1994). *Managing customer value: Creating quality and service that customers can see*. New York, NY: The Free Press.

80. Gallarza, M. G., & Saura, I. G. (2006). Value dimensions, perceived value, satisfaction and loyalty: An investigation of university students' travel behavior. *Tourism Management, 27,* 437–452.

81. Garcia-Dastugue, S. J., & Douglas, M. L. (2003). Internet-enabled coordination in the supply chain. *Industrial Marketing Management, 32,* 251–263.

82. Garvin, D. A. (1984). Product quality: An important strategic weapon. *Business Horizons, 27*(3), 40–43.

83. Garvin, D. A. (1984). What does product quality really mean? *MIT Sloan Management Review, 26*(1), 25–43.

84. Garvin, D. A. (1987). An agenda for research on the flexibility of manufacturing process. *International Journal of Operations and Production Management, 7*(1), 38–49.

85. Garvin, D. A. (1987). Competing on the eight dimensions of quality. *Harvard Business Review, 65*(6), 101–109.

86. Garvin, D. A. (1988). *Managing quality: The strategic and competitive edge*. New York, NY: Free Press.

87. George, J. M., & Jones, G. R. (2002). *Organizational behavior*. Upper Saddle River, NJ: Prentice Hall.

88. Gerstner, E. (1985). Do higher prices signal higher quality? *Journal of Marketing Research, 22*(5), 209–215.

89. Goldstein, E. B. (2010). *Sensation and perception* (10th ed.). Independence, KY: Cengage Learning.

90. Grant, R. M. (1996). Toward a knowledge-based theory of the firm. *Strategic Management Journal, 17,* 109–122.

91. Gribbin, J. R. (2003). *The scientists: A history of science told through the lives of its greatest inventors*, New York: Random House, Inc.

92. Grisaffe, D. B., & Kumar, A. (1998). Antecedents and consequences of customer value: Testing an expanded framework. *MSI Working Paper Series* No. 98–107.

93. Guest, D. E. (1987). Human resource management and industrial relations. *Journal of Management Studies, 24*(5), 503–521.
94. Gunter, B. (1989). The use and abuse of C_{pk}. *Quality Progress: Statistical Corner, 22*(1), 72–73.
95. Gupta, M., & Campbell, V. S. (1995). The cost of quality. *Production and Inventory Management Journal, 36*(3), 43–49.
96. Haapaniemi, P. (2001). Retooling the e-business strategy, *Chief Executive*, Nov, pp. 6–9.
97. Hackman, J., & Wageman, R. (1995). Total quality management: empirical, conceptual, and practical issues. *Administrative Science Quarterly, 40*, 309–342.
98. Handfield, R., & Nichols, E. (1999). *Introduction to supply chain management.* Upper Saddle River, NJ: Prentice-Hall Inc.
99. Harps, L. H. (2002, January). *Getting started in reverse logistics*, www.inboundlogistics.com.
100. Hedonism, (2004, April 20). Stanford Encyclopedia of Philosophy.
101. History of the Rickenbacker Motor Company. http://rickenbackermotors.com/hrm/hrm.html.
102. Horn, D., & Salvendy, G. (2006). Consumer-based assessment of product creativity: A review and reappraisal. *Human Factors and Ergonomics in Manufacturing, 16*(2), 155–175.
103. Houlihan, J. (1988). International supply chains: A new approach. *Quarterly Review of Management Technology, 26*(3), 13–19.
104. Hsu, Y., Pearn, W., & Wu, P. (2008). Capability adjustment for gamma processes with mean shift consideration in implementing six sigma program. *European Journal of Operational Research, 191*, 517–529.
105. Hughes, J. (1994). *Breaking the Tradition.* London: Wye College Press.
106. Hull, C. (1943). *Principles of behavior.* New York, NY: Appleton-Century-Crofts.
107. Inman, R. A., & Harry Hubler, J. (1992). Certify the process, not just the product. *Production and Inventory Management Journal, 33*(4), 11–14.
108. Ittner, C. D. (1986). Exploratory evidence on the behavior of quality costs. *Operations Research, 44*(1), 114–130.
109. Jarillo, J. C. (1993). *Strategic networks: Creating the borderless organization.* Oxford, UK: Butterworth Heinemann.
110. Juran, J. M., & Gyrna, F. A. (1951). *Quality control handbook.* New York, NY: McGraw-Hill.
111. Karnes, C. L., Sridharan, S. V., & Kanet, J. J. (1995). Measuring quality from the consumer's perspective: A methodology and its application. *International Journal of Production Economics, 39*, 215–225.
112. Keller, K. L., & Tybout, A. (2002). The principle of positioning. *Market Leader, 19*(Winter), 65.
113. Kenyon, G. N., & Sen, K. C. (2012). A model for assessing consumer perceptions of quality. *International Journal of Quality and Service Sciences, 4*(2), 175–188.
114. Kenyon, G. N., & Sen, K. C. (2012). Customer perceptions, dimensions of service quality, and service systems. In A. Juan, T. Daradoumis, M. Roca, S. Grasman, & J. Faulin (Eds.), *Decision making in service industries: A practical approach.* Boca Raton, FL: Taylor & Frances Group, CRC Press.
115. Kilmann, R. H. (1985). Corporate culture: Managing the intangible style of corporate life may be the key to avoiding stagnation. *Psychology Today, 19*(4), 62–68.
116. Kolarik, W. J. (1995). *Creating quality concepts, systems, strategies, and tools.* New York: McGraw-Hill.
117. Kolesar, P., Van Ryzin, G., & Cutler, W. (1998). Creating customer value through industrialized intimacy. *Strategy Management Competition, 12*, 2–12.
118. Kortge, G. D., & Okonkwo, P. A. (1993). Perceived value approach to pricing. *Industrial Marketing Management, 22*(2), 133–140.
119. Kushler, R., & Hurley, P. (1992). Confidence bounds for capability indices. *Journal of Quality Technology, 24*, 188–195.

120. LaBarre, P. (1995). The seamless enterprise. *Industry Week, 244*(12), 22–34.

227. van Laarhoven, P., Berglund, M., & Peters, M. (2000). Third-party logistics in Europe—five years later. *International Journal of Physical Distribution and Logistics Management, 30*(5), 425–442.

122. Lagrosen, Y., & Lagrosen, S. (2005). The effects of quality management: A survey of Swedish quality professionals. *International Journal of Operations and Production Management, 25*(90), 940–952.

123. Lambert, D. M. (2010). Customer relationship management as a business process. *Journal of Business and Industrial Marketing, 25*(1), 4–17.

124. Lambert, D. M., Cooper, M. C., & Pagh, J. D. (1998). Supply chain management: Implementation issues and research opportunities. *International Journal of Logistics Management, 9*(2), 1–19.

125. Lamming, R. C. (1993). *Beyond partnership: Strategies for innovation and lean supply.* Hemel, Hempstead: Prentice-Hall Inc.

126. Langley, C. J., & Holcomb, M. C. (1992). Creating logistics customer value. *Journal of Business Logistics, 13*(2), 1–27.

127. Lapierre, J. (2000). Customer-perceived value in industrial contexts. *Journal of Business and Industrial Marketing, 15*(2/3), 122–140.

128. Larson, P. D., & Rodgers, D. S. (1998). Supply chain management: Definition, growth, and approaches. *Journal of Marketing Theory and Practice, 6*(4), 1–5.

129. Lee, H., & Billington, C. (1995). The evolution of supply chain management models and practice at Hewlett-Packard. *Interfaces, 25*(5), 42–63.

130. Lee, H. L., Padmanabhan, V., & Whang, S. (1997). Information distortion in a supply chain: The bullwhip effect. *Management Science, 43*(4), 546–558.

131. Levy, M., & Grewal, D. (2000). Guest editors' overview of the issue supply chain management in a networked economy. *Journal of Retailing, 76*(4), 415–429.

132. Lewin, K. (1935). *A dynamic theory of personality.* New York, NY: McGraw-Hill.

133. Lewin, K. (1951). In D. Cartwright (Ed.), *Field theory in social science: Selected theoretical papers.* New York, NY: Harper & Row.

134. Lewis, I., & Talalayevsky, A. (1997). Logistics and information technology: A coordination perspective. *Journal of Business Logistics, 16*(1), 141–157.

135. Lewis, C. I. (1929). *Mind and the world order: An outline of a theory of knowledge.* Mineola, NY: Charles Scribner's Sons (Reprinted in 1956 by Dover Publishing, New York, NY).

136. Li, M. (2009). The customer value strategy in the competitiveness of companies. *International Journal of Business and Management, 4*(2), 136–141.

137. Lotfi, Z., Sahran, S., & Mukhta, M. (2013). A product quality–supply chain integration framework. *Journal of Applied Sciences, 13*, 36–48.

138. Louvieris, P., Van Westering, J., & Driver, J. (2003). Developing an eBusiness strategy to achieve consumer loyalty through electronic channels. *International Journal of Wine Marketing, 15*(1), 44–53.

139. Mack, A., & Rock, I. (1998). *Intentional blindness.* Cambridge, MA: MIT Press.

140. Margavio, R. L., Margavio, T. M., & Fink, G. W. (1993). Quality improvement technology using the taguchi method. *The CPA Journal, 72*–75.

141. McAdam, R., & Bannister, A. (2001). Business performance measurement and change management within a TQM framework. *International Journal of Operations & Production Management, 21*(1), 88–108.

142. McClelland, D. C. (1961). *The achieving society.* Princeton, NJ: Van Nostrand.

143. McGinnis, M., & Vallopra, R. (1999). Purchasing and supplier involvement in process improvement: A source of competitive advantage. *Journal of Supply Chain Management, 35*(4), 42–50.

144. Mejza, M. C., & Wisner, J. D. (2001). The scope and span of supply chain management. *International Journal of Logistics Management, 12*(2), 37–55.

145. Mentzer, J. T., DeWitt, W., Keebler, J. S., Min, S., Nix, N. W., Smith, C. D., et al. (2001). Defining supply chain management. *Journal of Business Logistics, 22*(2), 1–25.

146. Mentzer, J. T., Flint, D. J., & Hult, T. M. (2001). Logistics service quality as a segment-customized process. *Journal of Marketing, 65*(October), 82–104.

147. Miles, R., & Snow, C. (1994). *Fit, failure, and the hall of fame: How companies succeed or fail.* New York, NY: The Free Press.

148. Miles, R. E., Snow, C. C., Meyer, A. D., & Coleman, H. J. (1978). Organizational strategy, structure, and process. *Academy of Management Review, 3*(3), 546–562.

149. Min, S., & Mentzer, J. T. (2000). The role of marketing in supply chain management. *International Journal of Physical Distribution & Logistics Management, 30*(9), 765–787.

150. Moberg, C. R., Seph, T. W., & Freese, T. L. (2003). SCM: Making the vision a reality. *Supply Chain Management Review, 7*(5), 34–39.

151. Mokyr, J. (1990). *The lever of riches: Technological creativity and economic progress.* New York, NY: Oxford University Press.

152. Monezka, R., Trent, R., & Handfield, R. (2002). *Purchasing and supply chain management* (2nd ed.). Cincinnati, OH: Southwestern Thomson Learning.

153. Mueller, D. C., & Raunig, B. (1999). Heterogeneities within industries and structure-performance models. *Review of Industrial Organization, 15*(4), 303–320.

154. Murthy, D. N. P., & Kumar, K. R. (2000). Total product quality. *International Journal of Production Economics, 67,* 253–267.

155. Narayanan, V. G., Raman, A. (2000, April). *Aligning incentives for supply chain efficiency.* Boston: Harvard Business School Publishing. (Case # 9–600–110).

156. Nelson, P. (1970). Information and consumer behavior. *Journal of Political Economy, 78*(2), 311–329.

157. Neureuther, B. D., & Kenyon, G. N. (2004). Quality improvement under budgetary and life-cycle constraints. *Quality Management Journal, 11*(2), 21–32.

158. Neureuther, B. D., & Kenyon, G. N. (2009). Mitigating supply chain vulnerability. *Journal of Marketing Channels, 16*(3), 1–19.

159. Nilson, T. H. (1992). *Value-added marketing: Marketing management for superior results.* Berkshire, UK: McGraw-Hill.

160. Nisbett, R. E. (2003). *The geography of thought: How Asians and westerners think differently ... and why.* New York, NY: Free Press.

161. Nitschke, T., & O'Keefe, M. (1997). Managing the linkage with primary producers: Experiences in the Australian grain industry. *Supply Chain Management, 2*(1), 4–6.

162. Norman, D. A. (2004). *Emotional design: Why we love (or hate) everyday things.* New York, NY: Basic Books.

163. Odlyzko, A. (2000). The history of communications and its implications for the internet. *White Paper, AT&T Labs—Research.*

164. Oh, H. (2003). Price fairness and its asymmetric effects on overall price, quality, and value judgments: The case of an upscale hotel. *Tourism Management, 24,* 397–399.

165. Oliver, R. (1974). Expectancy theory predictions of salesmen's performance. *Journal of Marketing Research, 11,* 243–253.

166. Olson, E. M., Slater, S. F., & Hult, G. T. M. (2005). The importance of structure and process to strategic implementation. *Business Horizons, 48*(1), 47–54.

167. Olson, E. M., Slater, S. F., & Hult, G. T. M. (2005). The Performance Implications of fit among business strategy, marketing organization structure, and strategic behavior. *Journal of Marketing, 69*(3), 49–65.

168. Ormanidhi, O., & Stringa, O. (2008). Porter's model of generic competitive strategies: An insightful and convenient approach to firms' analysis. *Business Economics, 43*(3), 55–64.

169. O'Conner, R. (2004). Secrets of the masters. *Supply Chain Management Review, 8*(1), 42–50.

170. O'Malley, P. (1998). Value creation and business success. *The Systems Thinker, 9*(2), 110–135.

171. Palmer, J. W., & Griffith, D. A. (1998). Information intensity: A paradigm for understanding web site design. *Journal of Marketing Theory and Practice, 6*(3), 38–42.
172. Papatheodorou, Y. (2003). The economic case for systems integration. *Industrial Management, 45*(3), 8–11.
173. Parasuraman, A., Zeithaml, V. A., & Berry, L. L. (1985). A conceptual model of service quality and its implications for future research. *Journal of Marketing, 49*(4), 41–50.
174. Pearn, W., & Wu, C. (2006). Production quality and yield assurance for processes with multiple independent characteristics. *European Journal of Operational Research, 173*, 637–647.
175. Peteraf, M. A. (1993). The cornerstones of competitive advantage: A resource-based view. *Strategic Management Journal, 14*, 179–191.
176. Peypoch, R. (1998). The case for electronic business communities. *Business Horizons, 4*(16), 17–20.
177. Pirsig, R. M. (1974). *Zen and the art of motorcycle maintenance: An inquiry into values*. Morrow, NY: HarperCollins Publishers.
178. Pitts, R. E, Jr. & Woodside, A. D. (1986). Personal values and travel decisions. *Journal of Travel Research, 22*, 20–25.
179. Porter, M. E. (1980). *Competitive strategy: Techniques for analyzing industries and companies*. New York, NY: The Free Press.
180. Porter, M. E. (1985). *Competitive advantage: Creating and sustaining superior performance*. New York, NY: The Free Press.
181. Porter, M. E. (1987). Managing value from competitive advantage to corporate strategy. *Harvard Business Review, 65*(3), 43–59.
182. Prajogo, D. I., & Sohal, A. S. (2006). The relationship between organizational strategy, total quality management (TQM), and organization performance: The mediating role of TQM. *European Journal of Operational Research, 168*, 35–50.
183. Prakash, V. (1984). Personal values and product expectations. In R. E. Pitts & A. D. Woodside (Eds.), *Personal values and consumer psychology*. Toronto, Canada: Lexington Books.
184. Prasad, S., & Tata, J. (2003). The role of socio-cultural, political-legal, economic, and educational dimensions in quality management. *International Journal of Operations and Production Management, 23*(5), 487–521.
185. Pyzdek, T. (1992). Process capability analysis using personal computers. *Quality Engineering, 4*(3), 419–440.
186. Quinn, R. E., Hildebrandt, H. W., Rogers, P. S., & Thompson, M. P. (1991). A competing values framework for analyzing presentational communications in management contexts. *Journal of Business Communications, 28*(3), 213–232.
187. Radstaak, B. G., & Ketelaar, M. H. (1998). *Worldwide logistics: The future of supply chain services*. Hague, The Netherlands: Holland International Distribution Council.
188. Rayner, S. R. (1993). Metaphysical quality: The next quality battleground. *Non-Profit Management Strategies*, Rayner & Associates, Inc., Newletter.
189. Reeves, C. A., & Bednar, D. A. (1994). Defining quality: Alternatives and implications. *Academy of Management Review, 19*(3), 419–445.
190. Rich, N., & Holweg, M. (2000). *Value analysis value engineering*. Cardiff, UK: InnoRegio Project, Lean Enterprise Research Centre.
191. Riley, B. S., & Li, X. (2011). Quality by design and process analytical technology for sterile products: Where are we now? *AAPS PharmSciTech, 12*(1), 114–118.
192. Rimm, R., & Media, D. (2013). *Inventory and customer satisfaction*, Houston Chronicle, http://smallbusiness.chron.com/inventory-customersatisfaction-14324.html.
193. Robbins, S. P., & Judge, T. A. (1984). *Essentials of organizational behavior*. Upper Saddle River, NJ: Prentice Hall.
194. Rosan, M. A., & Kishawy, H. A. (2012). Sustainable manufacturing and design: Concepts, practices, and needs. *Sustainability, 4*, 154–174.
195. Rudzki, R. (2001). The next step: E-procurement. *New Steel* (March), pp. 24–25.

196. Russell, R. S., & Taylor, B. W. (2003). *Operations management* (4th ed.). Upper Saddle River, NJ: Prentice-Hall Inc.
197. Saura, I. G., Frances, D. S., Contri, G. B., & Blasco, M. F. (2008). Logistics service quality: A new way to loyalty. *Industrial Management and Data Systems, 108*(5), 650–668.
198. Scannell, T., Vickery, S., & Dodge, C. (2000). Upstream supply chain management and competitive performance in the automotive industry. *Journal of Business Logistics, 21*(1), 23–48.
199. Schacter, D., Gilbert, D. T., & Wegner, D. M. (2011). *Introducing Psychology*. New York, NY: Worth Publishers.
200. Schmenner, R. W., & Tatikonda, M. V. (2005). Manufacturing process flexibility revisited. *International Journal of Operations and Production Management, 25*(12), 1183–1189.
201. Schonberger, R. J. (1992). Is strategy strategic? Impact of total quality management on strategy. *Academy of Management Executive, 6*(3), 80–87.
202. Schonenberg, H., Mans, R., Russell, N., Mulyar, N., van der Aalst, W. M. P. (2008). Process flexibility: A survey of contemporary approaches. In J. Dietz, A. Albani & J. Barjis (Eds.), *Advances in enterprise engineering I*, Lecture Notes in Business Information Processing, (Vol. 10, pp. 16–30). Berlin: Springer.
203. Schumpeter, J. A. (1943). *Capitalism, socialism, and democracy* (6th ed.). New York, NY: Routledge.
204. Schwartz, S. H. (1996). Value priorities and behavior: Applying a theory of integrated value systems. In L. Erlbaum (Ed.), *The Psychology of Values* 8. The Ontario Symposium.
205. Scott, J. E., & Lamont, L. H. (1973). Relating consumer values to consumer behavior: A model and method for investigation. In T. W. Green (Ed.), *Increasing marketing productivity and conceptual and methodological foundations of marketing, Series 35*. Chicago, IL: American Marketing Association.
206. Sen, K. C. (1998). The use of franchising as a growth strategy by US restaurant franchisors. *Journal of Consumer Marketing, 15*(4), 397–407.
207. Sergiovanni, T. J. (1982). Ten principles of quality leadership. *Educational Leadership, 39*(5), 330–336.
208. Shaw, A. (1912). Some problems in market distribution. *Quarterly Journal of Economics, 26*, 708–765.
209. Shepard, W. C. (1990). *The economics of industrial organization* (3rd ed.). Upper Saddle River, NJ: Prentice-Hall, Inc.
210. Shewhart, W. A. (1939). *Statistical method from the viewpoint of quality control.* Washington, DC: The Graduate School, The Department of Agriculture (Reprinted in 1986 by Dover Publishing, New York, NY).
211. Sila, I., & Ebrahimpour, M. (2005). Critical linkages among TQM factors and business results. *International Journal of Operations and Production Management, 25*(11), 1123–1155.
212. Silver, E. A., Pyke, D. F., & Peterson, R. (1998). *Inventory management and production planning and scheduling*. New York, NY: Wiley.
213. Simatupang, T. M., & Sridharan, R. (2002). The collaborative supply chain. *International Journal of Logistics Management, 13*(1), 15–30.
214. Simon, H. A. (1991). Bounded rationality and organizational learning. *Organization Science, 2*, 125–134.
215. Slack, N. (1987). The flexibility of manufacturing systems. *International Journal of Operations and Production Management, 7*(4), 35–45.
216. Slack, N., & Lewis, M. (2003). *Operations strategy*. Upper Saddle River, NJ: Prentice-Hall Inc.
217. Smallwood, D. E., & Conlisk, J. (1979). Product quality in markets where consumers are imperfectly informed. *Quarterly Journal of Economics, 93*, 1–23.
218. Smith, A. (1776). *An inquiry into the nature and causes of the wealth of nations*. New York, NY: Random House, 1937.
219. Smock, D. (2003). Supply chain management: What is it? *Purchasing, 132*(13), 45–49.

220. Snow, C. C., Miles, R. E., & Coleman, H. J. (1992). Managing 21st century network orga-nizations. *Organizational Dynamics, 20*(3), 5–20.
221. Stevenson, W. J. (2014). *Operations management* (12th ed.). New York, NY: McGraw-Hall Irwin.
222. Stock, J. R. (1997). Applying theories from other disciplines to logistics. *International Journal of Physical Distribution and Logistics Management, 27*(9), 515–539.
223. Stone-Remero, E., & Grewal, D. (1997). Development of a multi-dimensional measure of perceived product quality. *Journal of Quality Management, 2*(1), 87–111.
224. Sum, C. C., & Teo, C. B. (1999). Strategic posture of logistics service providers in Singapore. *International Journal of Physical Distribution and Logistics Management, 29*(9), 588–605.
225. Svensson, G. (2002). The theoretical foundation of supply chain management: A func-tionalist theory of marketing. *International Journal of Physical Distribution and Logistics Management, 32*(9), 734–754.
226. Tan, T., & Alp, O. (2009). An integrated approach to inventory and flexible capacity man-agement subject to fixed costs and non-stationary stochastic demand. *OR Spectrum, 31,* 337–360.
227. Taylor, F. W. (1911). *The orinciples of scientific management.* New York, NY: Harper and Row.
228. Vachon, S., & Klassen, R. D. (2002). An exploratory investigation of the effects of sup-ply chain complexity on delivery performance. *IEEE Transactions on Engineering Management, 49*(3), 218–230.
229. Vaxevandis, N. M., & Petropoulos, G. (2008). A literature survey of cost of quality mod-els. *Journal of Engineering, 6*(3), 274–283.
230. Vicere, A. A. (2000). New economy, new HR. *Employment Relations Today, 27*(3), 1–11.
231. Vickers, J. (2011). The problem of induction. In N. Z. Edward (Ed.), *The stanford ency-clopedia of philosophy* (Fall ed.). http://plato.stanford.edu/archives/fall2011/entries/induction-problem/.
232. Vinson, D. E., Scott, J. E., & Lamont, C. M. (1977). The role of personal values in mar-keting and consumer behavior. *Journal of Marketing, 41*(2), 44–50.
233. Vollman, T. E., Berry, W. L., & Whybark, D. C. (2005). *Manufacturing planning and con-trol systems* (5th ed.). New York, NY: Irwin/McGraw Hill Companies Inc.
234. Vroom, V. H. (1964). *Work and motivation.* New York, NY: Wiley.
235. Waldman, D. A. (1993). A theoretical consideration of leadership and total quality man-agement. *Leadership Quarterly, 4*(1), 65–79.
236. Walter, A., Muller, T. A., Helfert, G., & Ritter, T. (2003). Functions of industrial supplier relationships and their impact on relationship quality. *Industrial Marketing Management, 32*(2), 159–169.
237. Wauters, F., & Mahot, J. (2002). *OEE overall equipment effectiveness.* White Paper, ABB, Inc.
238. Weiten, W. (2008). *Psychology: Themes and variations.* Independence, KY: Cengage Learning.
239. Weld, L. D. H. (1916). *The marketing of farm products.* New York, NY: The MacMillan Company.
240. Whipple, J. M., & Frankel, R. (2000). Strategic alliance success factors. *Journal of Supply Chain Management, 36*(2), 21–28.
241. Wigfield, A., & Eccles, J. S. (1994). Children's competence beliefs, achievement values, and general self-esteem. *Journal of Early Adolescence, 14*(2), 107–139.
242. Wiley, C. (2013). *What is perceptual mapping in product development?* Demand media in Houston Chronicle.
243. Wilkinson, A. (1992). The other side of quality: Soft issues and the human resource dimension. *Total Quality Management, 3*(3), 323–329.

244. Williams, L. R., Esper, T. L., & Ozment, J. (2000). The electronic supply chain: Its impact on the current and future structure of strategic alliances, partnerships, and logistics leadership. *International Journal of Physical Distribution & Logistics Management, 32*(8), 703–718.

245. Womack, J. P., Jones, D. T., & Roos, D. (1990). *The machine that changed the world.* New York: Maxwell, Macmillan.

246. Wu, C. (2009). Decision-making in testing process performance with fuzzy data. *European Journal of Operational Research, 193,* 499–509.

247. Yang, Z., Jun, M., & Peterson, R. T. (2004). Measuring customer perceived online service quality. *International Journal of Operations and Production Management, 24*(11), 1149–1174.

248. Zeithaml, V. A. (1988). Consumer perceptions of price quality, and value: A means-end model and synthesis of evidence. *Journal of Marketing, 52,* 2–22.

249. Zu, X., Robbins, T. L., & Fredendall, L. D. (2010). Mapping the critical links between organizational culture and TQM/Six Sigma practices. *International Journal of Production Economic, 123,* 86–106.

Printed in the United States
By Bookmasters